Carbon-Capture by Metal-Organic Framework Materials

David J. Fisher

Published by **Materials Research Forum LLC**
Millersville, PA 17551, USA

Published as part of the book series
Materials Research Foundations
Volume 77 (2020)
ISSN 2471-8890 (Print)
ISSN 2471-8904 (Online)

Print ISBN 978-1-64490-084-0
ePDF ISBN 978-1-64490-085-7

This book contains information obtained from authentic and highly regarded sources. Reasonable efforts have been made to publish reliable data and information, but the author and publisher cannot assume responsibility for the validity of all materials or the consequences of their use. The authors and publishers have attempted to trace the copyright holders of all material reproduced in this publication and apologize to copyright holders if permission to publish in this form has not been obtained. If any copyright material has not been acknowledged please write and let us know so we may rectify in any future reprint.

Distributed worldwide by

Materials Research Forum LLC
105 Springdale Lane
Millersville, PA 17551
USA
https://www.mrforum.com

Printed in the United States of America
10 9 8 7 6 5 4 3 2 1

Table of Contents

Introduction

Global-warming is now recognized as being an existential threat to humanity, and one of the main contributory factors is the increase in so-called greenhouse gases due to human activity. The main offending gas is carbon dioxide, and so every means for reducing its incidence in the atmosphere must be explored. Among the materials which promise to be valuable in this regard are the present ones: the metal-organic frameworks, also known as coordination polymers. The present work covers the most recent advances made in this field.

They can help by capturing and storing carbon dioxide, or even by aiding the synthesis of the gas into useful materials. There are many familiar materials suitable for the absorption of carbon dioxide, such as activated carbon and certain zeolites, but they suffer from various problems. The advent of this new class of materials has greatly widened the available possibilities. Suitable solid adsorbents for CO_2 require a low total regeneration energy, and this can be arranged by combining a high working-capacity with a low desorption temperature and a narrow temperature-difference between capture and regeneration.

The metal-organic frameworks are inorganic metal ions or clusters of metal atoms, especially zinc, copper and iron - plus bridged organic ligands - which interconnect by self-assembly to form crystalline highly-porous solids having chemically stable periodic network structures. As for the ligands, the favourite choice is a stable carboxylic acid.

They offer many opportunities for investigating active sites for catalysis, and a common quest is to find a metal site with labile ligands that can be eliminated so as to create a coordinate-wise unsaturated site: a so-called open-metal site. Although they are common, the relationship between their atomic-level structure and catalytic properties is not yet clear, but the study of non-covalent interactions provides insights into the relationship between the catalytic performance and the framework structure.

The wide choice of metals and organic linkers as building-blocks for these materials means that there is a theoretically unlimited number of frameworks, together with the possibility of 'decorating' the pores with various functional groups, as well as a selection of organic ligands and inorganic building blocks.

As well as simply sequestering carbon dioxide, its direct electrochemical reduction into useful chemicals and fuel is one of the most promising approaches. This might seem to be counter-productive, since most fuels will simply release the gas again when burned. Widespread use of this form of carbon-capture would nevertheless reduce the need to newly mine fossil-fuels such as coal and would at least stabilise the current CO_2 burden.

Electrochemical reduction can, for example, be used to convert CO_2 into methanol by employing alternative energy sources and electrocatalysts. The choice of the latter is a key factor. Unadulterated metals, metal oxides, and now metal-organic frameworks, have been used but the latter now attract most interest. In addition to CO_2 reduction, metal-organic frameworks can aid electrocatalytic hydrogen evolution, thus boosting the 'hydrogen economy'; which also promises to take the pressure off fossil-fuel use. The challenge here is to identify electrocatalysts of high efficiency and selectivity for the electro-reduction of CO_2.

One of the main products of such reduction is CO; an apparently counter-productive step since the monoxide is a fuel which will soon revert to the dioxide. Recycling is nevertheless superior to the mining of new CO_2-emitting energy sources. Among the other possible products are formic acid, hydrocarbons … and the aforementioned methanol. The most efficient traditional catalysts usually involve expensive and scarce elements, but metal-organic frameworks generally contain relatively cheap components. Their unique properties often provide new ways of proceeding. For example, there is no standard technique for fixing a catalyst directly to the surface of a conductive substrate without using a polymeric adhesive, but a metal-organic hybrid carbon photo-electrocatalytic framework material has been be attached to the surface of a macroporous metal support by carbonization. A catalyst which was carbonized at 700C led to an up to 9 times higher methanol yield. Product selectivity could also be switched from methanol to CO by choosing between electrocatalysis and photocatalysis.

The problem of the 'pick and mix' nature of metal-organic frameworks, and the consequent existence of a very large number of candidate structures, has been alleviated by the use of high-throughput computational screening. One difficulty which has arisen is the somewhat haphazard naming of these framework materials. Straightforward listing of the metal-organic components leads to rather unwieldy names, but the distinguishing of them by using serial-numbers or trade-names has been uncoordinated and has led, for instance, to the same material sometimes having two different names, as in the case of Copper-BTC and HKUST-1. Several organisations have initiated their own named series, leading to the UiO metal-organic frameworks of the University of Oslo, the DUT series of Dresden University of Technology, the ZU series of Zhejiang University, The HNUST series of Hunan University of Science and Technology, the MIL materials of Institut Lavoisier, the NJU series of Nanjing University and the CFA series: Coordination Framework Augsburg University. This impedes computer-searching and the correlation of structures and properties. The *ad hoc* naming approach has also made it difficult to decide how best to list those materials in the current work. Given the arbitrariness of the above naming systems, the logical solution is here to list the materials in the order of the metallic component of the metal-organic framework.

Various methods are available for the preparation of metal-organic frameworks. In solvothermal synthesis, mixed solutions of an inorganic salt and an organic ligand are introduced into a sealed vessel and heated to as high as the boiling point in order to form an insoluble framework. In electrochemical synthesis, the metal salt may be directly introduced into an electrolyte solution, together with the organic ligand.

Porosity and a consequent large specific surface area are the principal features of these materials, thus making them very useful for gas adsorption, gas separation and catalysis; especially as the organic ligands and metal ion-clusters of the metal-organic frameworks form particularly regular spatial networks. Guest-molecules are involved in the growth of these crystals, and the pores are formed by their removal. The specific surface area is a key measure of adsorption and catalytic capabilities, and the latter tend to increase with increasing specific surface area. The constant developmental aim is therefore to choose organic ligands and connecting molecules so as to maximise the specific surface area. The usual parameter is that of Brunauer, Emmett and Teller (BET), as determined for a standard temperature and pressure, and typical values are shown in table 1.

Table 1. Brunauer-Emmett-Teller (BET) values of typical metal-organic frameworks

Name	Ligand	BET (m^2/g)
MOF-205	2,6-naphthalenedicarboxylate	4460
MOF-177	4,4',4"-benzene-1,3,5-triyltribenzoate	4500
MOF-200	4,4',4"-(benzene-1,3,5-triyl-tris (benzene-4,1-diyl)) tribenzoate	4530
MOF-210	biphenyl-4,4'-dicarboxylate	6240

The potential advantage of using supported ionic liquids for CO_2 capture was investigated[1] by introducing 1-ethyl-3-methylimidazolium acetate into the well-known voluminous metal–organic frameworks, MOF-177 and MIL-101. A marked decrease in porosity occurred as a result of modification, especially when using wet impregnation methods, but the crystal structures were retained. No improvement in CO_2 take-up was found for MIL-101, but the modified MOF-177 samples - prepared using wet impregnation – exhibited a marked increase in CO_2 capacity: up to 0.3mmol/g at 0.15bar and 303K. A microporous metal organic framework (MOF-177) was synthesized and used to investigate[2] the static and dynamic adsorption behavior of CO_2 and CH_4. The synthesized MOF-177 contained channels with an average pore diameter of 1.18nm. There existed distinct crystals having a needle-like geometrical shape containing large pores with a diameter of the order of 20.15Å. The surface area of MOF-177 was 1721m^2/g, with a CO_2 adsorption capacity of 1.03mmol/g and a CO_2/CH_4 equilibrium selectivity of 3.21 at 1atm and 25C. The MOF-177 exhibited a notable regeneration ability and maintained its CO_2 adsorption capacity over several adsorption-desorption cycles. The dynamic separation of binary mixtures having 2 compositions (CO_2:CH_4 30:70 and CO_2:CH_4 70:30) through a fixed-bed column indicated that CH_4 passed through MOF-177 faster than did CO_2; indicating a higher selectivity for CO_2 as compared with CH_4.

Methacrylamide has been polymerized to polymethacrylamide within MIL-101, and reduced with lithium aluminium hydride[3]. The resultant adsorbent materials had improved performances with regard to CO_2 capture and selectivity over N_2. The reduced polymethacrylamide material adsorbed 1.4mmol/g of CO_2 at 0.15atm and 298K, with a CO_2/N_2 selectivity of 143 at 298K and 1atm, with $CO_2/N_2 = 0.75/0.15$. These values were about 3 times and 10 times higher, respectively, than those for plain MIL-101. The adsorbent could be easily recycled.

A study was made of the adsorption performance of metal-organic frameworks impregnated with ionic liquids[4]. The adsorption and diffusion of light gases (CO_2, CH_4, N_2) and their mixtures in hybrid composites were calculated using molecular simulations. The hybrid composites consisted of 1-ethyl-3-methylimidazolium thiocyanate impregnated in IRMOF-1, HMOF-1, MIL-47, and MOF-1. The increase in the amount of ionic liquid enhanced the adsorption selectivity in favour of carbon dioxide for the mixtures CO_2/CH_4 and CO_2/N_2 and in favour of methane in the mixture, CH_4/N_2.

The adsorption of CO_2 on metal-organic frameworks is a physical process in which the adsorption sites and guest molecules form stable structures via electrostatic and other forces. The adsorption is governed mainly by the specific surface area and the nature of the functional groups. In the presence of multi-component gases, CO_2 adsorption depends mainly upon pore-shape, pore-size, temperature and pressure. The CO_2 adsorption and separation capabilities can be enhanced by changing the organic ligand and adding unsaturated metal sites. Under especially high pressures, the CO_2 adsorption is determined mainly by the specific surface area and void volume of the material, with those having a greater specific surface area and void volume tending to exhibit a higher CO_2 saturation limit on adsorption capacity. At low pressures, materials having a large surface area exhibit poor CO_2-trapping effect. This is due mainly to the adsorption heat of the material.

Materials Research Forum LLC
https://doi.org/10.21741/9781644900857

Electrochemical reduction of CO_2 using renewable energy is one of the most promising approaches to CO_2 conversion, with reaction usually performed in an electrochemical cell having 2 compartments which are separated by a Nafion film. Water in an electrolyte acts as the proton source and reduction occurs at the cathode, together with oxygen evolution at the anode. Electrocatalytic CO_2 reduction[5] can involve a number of reactions:

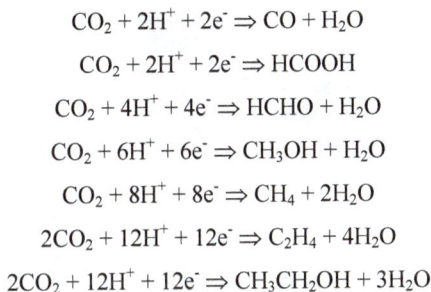

$$CO_2 + 2H^+ + 2e^- \Rightarrow CO + H_2O$$

$$CO_2 + 2H^+ + 2e^- \Rightarrow HCOOH$$

$$CO_2 + 4H^+ + 4e^- \Rightarrow HCHO + H_2O$$

$$CO_2 + 6H^+ + 6e^- \Rightarrow CH_3OH + H_2O$$

$$CO_2 + 8H^+ + 8e^- \Rightarrow CH_4 + 2H_2O$$

$$2CO_2 + 12H^+ + 12e^- \Rightarrow C_2H_4 + 4H_2O$$

$$2CO_2 + 12H^+ + 12e^- \Rightarrow CH_3CH_2OH + 3H_2O$$

with the resultant product depending upon the activation energy (table 2). One strategy for using a metal-organic framework as the catalyst has been to embed copper nanoparticles in thin-film zirconium-based framework material, NU-1000, by incorporating single Cu^{II} sites via solvothermal deposition; thus influencing the production-rate of various products (figure 1).

Table 2. Activation energies for electrochemical reduction products of CO_2

Product	Energy (eV)
HCOOH	-0.250V
CO	-0.106V
HCHO	-0.070V
CH_3OH	0.016V
C_2H_4	0.064V
CH_3CH_2OH	0.084V
CH_4	0.169V

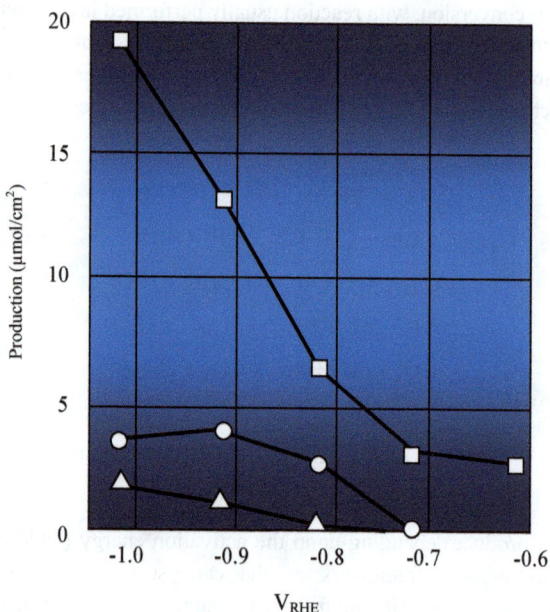

Figure 1. Production rates of the products
of Cu-SIM NU-1000 thin film catalysts
squares: H_2, circles: HCOO, triangles: CO

Aluminium-

Hybrids have been produced[6] which comprised silver nanocrystals and aluminium-based porphyrin metal-organic frameworks: ([$Al_2(OH)_2(TCPP)$]), where TCPP is tetrakis(4-carboxyphenyl)porphyrin. The nanocrystals were embedded in the framework material while retaining electrical contact with a conductive substrate. This permitted the hybrid to be used as an electrocatalyst for CO_2 reduction. The clean interface between the nanocrystals and the framework material allowed for electronic changes, in the silver, which suppressed the hydrogen evolution reaction and promoted the reduction reaction. A photocatalyst comprising a porphyrin-based metal-organic framework and amine-functionalized graphene has also been designed[7] to reduce CO_2. It markedly improved

CO_2 photoreduction, as compared with an aluminium-based porphyrin metal-organic framework. This type of visible-light driven photocatalyst converted CO_2 to HCOO- at a formate evolution rate of 685.6μmol/gcath on an aluminium-based porphyrin metal-organic framework with 5wt% of amine-functionalized graphene.

A molecular catalyst and a molecular photosensitizer were co-immobilized[8] in the highly porous material, MIL-101-NH$_2$(Al). The resultant composite permitted the reduction of CO_2 in visible light, and exhibited marked specificity; with the exclusive production of formate.

Metal-organic framework crystal-glass composites are materials in which a crystalline framework material is dispersed within a metal-organic framework glass. The room-temperature stabilization of the open-pore form of MIL-53(Al), usually observed at high temperatures, which occurs upon encapsulation within a ZIF-62(Zn) metal-organic glass matrix was explored[9]. A series of framework crystal-glass composites containing different loadings of MIL-53(Al) were synthesized, and characterized using X-ray diffraction and nuclear magnetic resonance spectroscopy. An upper limit of MIL-53(Al) that could be stabilized in the composite was determined for the first time.

Heat-treatment, PDMS-coating and combined thermal treatment with PDMS coating methods were applied to NH$_2$-MIL-53(Al)/cellulose acetate hollow-fiber mixed-matrix membranes (HFMMM) for CO_2/CH_4 and CO_2/N_2 separation. The results showed[10] that upon thermal treatment at 170C, CO_2/CH_4 and CO_2/N_2 gas pair selectivities of HFMMM were improved from 9.1 to 9.9 and 7.7 to 8.4, respectively. On the other hand, CO_2/CH_4 and CO_2/N_2 gas pair selectivities of HFMMMs were improved from 9.1 to 14.1 and 7.7 to 12.0, respectively, after heat treatment at 220C. Substantial increments in the gas pair selectivities of the thermally treated fibers were due mainly to polymer densification, which condensed the voids between polymer and particles. PDMS-coated HFMMMs subsequently exhibited gas pair selectivities of 17.0 and 16.2, respectively, which were higher in comparison to the untreated and thermally treated HFMMM. The highest improvements in CO_2/CH_4 and CO_2/N_2 gas pair selectivities were found for combined post-treated HFMMM with selectivities values of 26.6 and 24.9, respectively. This result was due to the combined effect of polymer densification and the penetration of PDMS solution into the cellulose acetate matrix, which contributed to the suppression of non-selective voids.

The successful chloro-functionalization of aluminium fumarate (MIL-53-Fum) was achieved by *in situ* hydrochlorination of acetylenedicarboxylic acid upon reaction with aluminium chloride, resulting[11] in the formation of the aluminium chlorofumarate metal-organic framework (MIL-53-Fum-Cl = [Al(OH)(Fum-Cl)]) in a one-pot reaction. The chloro functional groups decorating the pores enhanced gas (CO_2, CH_4, H_2) sorption capacities and affinity as compared with non-functionalized MIL-53-Fum. The functionalization also resulted in a 2-fold increase in the selective adsorption of CO_2 over CH_4 as compared with MIL-53-Fum.

The construction of membranes ultimately decides the performance of the membrane, with the underlying interaction between fillers and polymer playing the most important role concerning pathways for gas molecules. The interactions between the metal–organic framework MIL-53(Al) and polysulfone in MMMs were compared[12] with various concentrations of polysulfone in two different casting solvents, chloroform and tetrahydrofuran. The preparation of membranes with varying degrees of sedimentation, subsequent analysis, and gas permeation performance was described. The morphology of MMMs thus prepared exhibited filler sedimentation at low precursor solution concentrations, which was characterized by scanning electron microscopy with the positions of filler further detected by energy-dispersive X-ray spectroscopy. The results of X-ray diffraction confirmed an obvious discrepancy of intensity caused by sedimentation. An investigation was made of the relationship between viscosity and thermal analysis, measured by rheometer, differential scanning calorimetry and thermogravimetric analysis, where the highest T_g and best thermal stability in a moderate viscosity of 60 to 70Pas also matched an homogeneous morphology having a stronger interaction between fillers and polymer. Further gas permeation measurements showed a promising CO_2/N_2 selectivity of 22.71 for an homogeneous membrane. The permeation properties of various preparations were also studied, revealing an influence of filler distribution upon gas permeation, related to different diffusion paths for gas molecules in sedimentation, aggregation, and homogeneous morphologies.

A novel 10%Co/MIL-53(Al) nanohybrid based upon the microporous metal-organic framework, MIL-53(Al), prepared by MW-assisted synthesis exhibits[13] the properties of a bifunctional catalyst active in CO_2 hydrogenation to CO and in the consecutive synthesis of liquid hydrocarbons from the carbon monoxide formed. An observed increase in the CO_2 conversion, in comparison with the thermodynamically possible conversion, is assumed to be due to the expected shift of the equilibrium towards the formation of CO as a result of its further rapid conversion to hydrocarbons by the Fisher-Tropsch reaction.

Rapidly (25C, 60s) synthesised metal-organic framework, aluminium fumarate (AlFu), was impregnated with different amounts of tetraethylenepentamine to build novel CO_2 adsorbents[14]. The results showed that the introduction of tetraethylenepentamine into AlFu was successfully achieved by impregnation. The CO_2 adsorption performances of AlFu and TEPA-AlFu were investigated by the TGA method. The effect of adsorption temperature, tetraethylenepentamine loading and CO_2 partial pressure upon the adsorption performance was studied. The results showed that the optimum adsorption temperature of TEPA-AlFu was 75C, and that increasing the loading amount of tetraethylenepentamine increased the CO_2 adsorption capacity and decreased the adsorption rate. A high CO_2 partial pressure was good for the adsorption process, and the highest CO_2 adsorption capacity of 4.10mmol/g was obtained at 75C in a pure CO_2 atmosphere, for AlFu loaded with 60wt% tetraethylenepentamine. Studies of the adsorption kinetics showed that the fractional-order kinetic model fitted the CO_2 adsorption experiment data well. The CO_2 adsorption capacity of 60TEPA-AlFu sorbent decreased by only 2.81% after 10 adsorption-desorption cycles performed within the temperature range of CO_2 adsorption, 75C, and desorption: 110C.

The same metal-organic framework filler, in porous and non-porous states, was compared so that artefacts arising from a different polymer-filler interface were factored-out. Mixed-matrix membranes having the porous metal-organic framework, aluminium fumarate (Al-fum) and with a non-porous dimethyl sulphoxide solvent-filled aluminium fumarate, Al-fum(DMSO), both with Matrimid as polymer, were prepared[15]. The filler contents ranged from 4 to 24wt%. The gas-separation performances of both membranes were determined using mixed-gas measurements, with a binary mixture of CO_2/CH_4 and gas permeation following the theoretical predictions of the Maxwell model for both a porous and non-porous dispersed phase (filler). Membranes with the porous Al-fum filler showed an increased CO_2 and CH_4 permeability, with a moderate rise in selectivity upon increasing the filler fraction. The membranes with non-porous Al-fum(DMSO) filler underwent a reduction in permeability, while maintaining the selectivity of the neat polymer. A linear dependence of log[P] versus the reciprocal specific free fractional volume (sFFV) ruled out any significant contribution arising from void volume. The sFFV included the free volume of the polymer and the metal-organic framework, but not the polymer-filler interface volume (so-called void volume). The sFFV for the membrane was calculated to be between $0.23cm^3/g$ for a 24wt%Al-fum/Matrimid MMM and $0.12cm^3/g$ for a 24wt%Al-fum(DMSO)/Matrimid membrane. The negligible effect of an interface volume is supported by a good matching of theoretical and experimental density of the Al-fum and Al-fum/(DMSO) membranes which gave a specific void volume below $0.02cm^3/g$; and often even below $0.01cm^3/g$.

Carboxylic acid linker ligands are known to form strong metal-carboxylate bonds to afford many different variations of permanently microporous metal-organic frameworks. A controlled approach to decarboxylation of the ligands in carboxylate-based metal-organic frameworks could result in structural modifications, offering scope to improve existing properties or to unlock entirely new properties. It was demonstrated[16] that the microporous metal-organic framework, MIL-121, is transformed to a hierarchically porous metal-organic framework via thermally triggered decarboxylation of its linker. Decarboxylation and the introduction of hierarchical porosity increases the surface area of this material from 13 to $908m^2/g$ and enhances gas adsorption take-up for industrially relevant gases (i.e., CO_2, C_2H_2, C_2H_4, and CH_4). For example, CO_2 take-up in hierarchically porous MIL-121 is improved 8.5 times over MIL-121, reaching $215.7cm^3/g$ at 195K and 1bar; CH_4 take-up is $132.3cm^3/g$ at 298K and 80bar in hierarchically porous MIL-121 versus zero in unmodified MIL-121. The approach taken was validated using a related aluminium-based metal-organic framework, ISOMIL-53. Loading metal guests within metal-organic frameworks via secondary functional groups is a promising route for introducing or enhancing metal-organic framework performance in various applications. Metal ions (Li^+, Na^+, K^+, Mg^{2+}, Ca^{2+}, Ba^{2+}, Zn^{2+}, Co^{2+}, Mn^{2+}, Ag^+, Cd^{2+}, La^{3+}, In^{3+} and Pb^{2+}) have been successfully introduced into MIL-121 metal-organic frameworks by using a cost-efficient route involving free carboxylic groups on the linker. The local and long-range structure of the metal-loaded frameworks was characterized[17] by using multinuclear solid-state nuclear magnetic resonance and X-ray diffraction methods. Li/Mg/Ca-loaded MIL-121 and silver nanoparticle-loaded MIL-121 exhibited enhanced H_2 and CO_2 adsorptions; silver nanoparticle-loaded MIL-121 also exhibited remarkable catalytic activity in the reduction of 4-nitrophenol.

Some 3p-block metal-organic frameworks, SNNU-5-Al, Ga and In, exhibited thermal stability[18] up to 500C, with SNNU-5-Al having an excellent water-tolerance (24 days at room temperature and 24h at 80C in pure water) and a pH stability of 4 to 12. The super-large cuboctahedral and octahedral cage structures imparted an excellent CO_2 take-up and separation ability. The strong Lewis-acid nature of the open 3p metal sites made these materials ideal heterogeneous catalysts for CO_2 catalytic fixation using epoxides.

The aluminium trimesate-based MOF, MIL-96-(Al), possesses a high chemical stability and a marked CO_2 adsorption ability. The CO_2 capture ability and selectivity has been further improved[19] by incorporating a second metal, calcium, and varying the Ca^{2+}/Al^{3+} molar ratio. This resulted in marked changes in crystal shape and size, with the shape changing from hexagonal rods capped with hexagonal pyramids, in the absence of calcium, to thin hexagonal disks at the highest calcium content. The CO_2 adsorption at pressures of up to 950kPa was greatly improved, due to increased pore volumes. At 100 and 28.8kPa, the CO_2/N_2 selectivity of the highest calcium-content variant attained 67 and 841.42, respectively; 5 and 26 times the selectivity of the calcium-free material.

Barium-

One novel metal-organic framework, $[Ba(L)(H_2O)_{1.5}]_n$ (H_2L = aniline-2,5-disulfonic acid), has been synthesized[20] by using a hydrothermal method. Each barium atom is eleven-coordinated into a distorted monocapped pentagonal antiprismatic arrangement. The compound exhibits an interesting 3-dimensional pillar-layered structure constructed from 2-dimensional inorganic layers $[Ba(SO_3)_2(H_2O)_{1.5}]_n$ and organic pillars of phenyl moieties of L2- linkages. The inorganic layers are supported by the organic pillars, generating a novel 3-dimensional open framework structure with {3, 4 6 , 5 5 , 6 5 , 7 4 } 2 {3}{5} topology. The results of fluorescence measurements reveal that the decayed emission band centered at 492nm may be caused by the interactions of the ligands and the metal ions. The $[Ba(L)(H_2O)_{1.5}]_n$ was selective towards the adsorption of CO_2 over N_2 at 273K.

An oxalamide-functionalized ligand, N,N'-bis(isophthalic acid)-oxalamide (H_4BDPO), was designed[21] such that solvothermal reaction of H_4BDPO with the oxophilic alkaline-earth, Ba^{2+} ion, created a honeycomb barium-based metal-organic framework, $\{[Ba_2(BDPO)(H_2O)]\cdot DMA\}_n$. Due to the existence of Lewis basic oxalamide groups and unsaturated Lewis acid metal sites in the tubular channels, the activated framework exhibited not only high C_2H_6, C_2H_4 and CO_2 take-ups and selective capture from CH_4, but also efficient CO_2 chemical fixation.

Bismuth-

Exclusive Bi-N4 sites on porous carbon networks can be obtained via the thermal decomposition of a bismuth-based metal-organic framework and dicyandiamide for CO_2RR. Interestingly, *in situ* environmental transmission electron microscopy analysis not only showed directly the reduction from Bi-MOF to bismuth nanoparticles but also exhibited a subsequent atomization of bismuth nanoparticles, assisted by the NH_3 released from the decomposition of dicyandiamide. The catalyst exhibits[22] a high intrinsic CO_2 reduction activity for CO conversion, with a high faradaic efficiency (up to 97%) and a high turnover frequency of 5535/h at a low overpotential of $0.39V_{RHE}$. Further experiments and density functional theory results demonstrate that the single-atom Bi-N4 site is simultaneously the predominant active center for CO_2 activation and for the rapid formation of the key intermediate COOH with a low free energy barrier.

Cadmium-

A 3-dimensional anionic metal-organic framework, $(Me_2NH_2)_2[Cd(PTC)]\bullet 2H_2O$, assembled from a Cd^{II} salt and a planar aromatic core ligand containing tetracarboxylate groups (H_4PTC = pyrene-1,3,6,8-tetracarboxylic acid), has been constructed[23] using a solvothermal method. In the structure of the fabricated Cd-PTC, each Cd^{II} center-links 4 carboxyl groups, and each PTC4- ligand connects four Cd^{II} ions through the de-protonated carboxylate groups in bi-dentate chelating style, generating a 3-dimensional network with 1-dimensional rhombic channels. This topology was considered to be a 3-dimensional (4,4)-connected pts net. Gas sorption measurements revealed that de-solvated Cd-PTC underwent a strong interaction with CO_2 and exhibited a highly selective CO_2 capture among the mixed gases, CO_2/CH_4 and CO_2/N_2.

Table 3. Thermal expansion coefficients of isostructural
$[M(pba)_2]\bullet2DMA$ metal-organic frameworks

Material	r_{ion} (Å)	α_a (/K)	α_b (/K)	α_b (/K)	α_V (/K)
$[Ni(pba)_2]\bullet2DMA$	0.69	1.53×10^{-4}	0.41×10^{-4}	-0.35×10^{-4}	1.59×10^{-4}
$[Co(pba)_2]\bullet2DMA$	0.72	1.93×10^{-4}	0.64×10^{-4}	-0.76×10^{-4}	1.80×10^{-4}
$[Zn(pba)_2]\bullet2DMA$	0.74	1.87×10^{-4}	0.89×10^{-4}	-0.84×10^{-4}	1.91×10^{-4}
$[Zn_{0.77}Cd_{0.23}(pba)_2]\bullet2DMA$	0.74-0.96	1.88×10^{-4}	1.42×10^{-4}	-1.34×10^{-4}	1.95×10^{-4}
$[Cd(pba)_2]\bullet2DMA$	0.96	2.26×10^{-4}	1.65×10^{-4}	-1.55×10^{-4}	2.33×10^{-4}

Solvothermal reactions of Hpba [3-(4-pyridyl)-benzoic acid] with a series of transition-metal ions produced[24] isostructural metal-organic frameworks, $[M(pba)_2]\bullet2DMA$, where M was Ni^{2+}, Co^{2+}, Zn^{2+}, Cd^{2+} or Zn^{2+}/Cd^{2+} and DMA was N,N-dimethylacetamide. These materials were 2-dimensional fence-like coordination networks, which were based upon mononuclear 4-connected metal nodes and 2-connected organic ligands, and exhibited positive and negative thermal expansions that could be greater than 1.50×10^{-4}/K: α_a = 1.54 to 2.28×10^{-4}/K, α_b = 0.41 to 1.64×10^{-4}/K and α_c = -0.37 to -1.52×10^{-4}/K. Larger metal ions led to larger expansion coefficients (table 3, figure 2) because the greater space led to increased ligand vibrational motion and hinged-fence effect, and also permitted greater changes in steric hindrance to occur between the layers. The $[Cd(pba)_2]\bullet2DMA$, in particular, had very large thermal expansion coefficients. Between 112 and 300K, its a-, b- and c-axes changed by 4.2, 3.1 and -2.9%, respectively; giving linear thermal expansion coefficients of α_a = 2.26×10^{-4}/K, α_b = 1.65×10^{-4}/K and α_c = -1.55×10^{-4}/K. The overall volume increase was 4.3%; corresponding to a thermal expansion coefficient of 2.33×10^{-4}/K. There was also a change in the internal angle of the metal-organic fences (table 4). By considering the Cd^{2+} ions to be 4-connected nodes and the pba ligands to be linkers, the coordination network could be viewed simply as a 2-dimensional rhombus grid or as a hinged fence with the 4-connected sql topology lying parallel to the bc-plane. Such 2-dimensional grids stack in a zig-zag offset fashion, via C-H$\cdots$O hydrogen bonding and C-H$\cdots$p interactions, to form 3-dimensional supramolecular structures. Thanks to the length of the ligands, 1-dimensional rhombus nano-channels with a cross-section of up to 8.5Å x 11.3Å form along the a-axis following layer-stacking, and the solvent-accessible void-size attains 46%. The N,N-dimethylacetamide components are closely packed, face-to-face, in the channels. Each metal ion is coordinated with 2 carboxyl groups and 2 pyridyl groups, arranged in a tetrahedral configuration, and the tetrahedral building units of adjacent layers intercalate one another. Larger metal ions produced more open tetrahedra (O2$\cdots$C24 and O4$\cdots$C12 between the pyridyl and carboxyl groups), less steric hindrance and smaller interlayer separations. Larger metal ions also allowed the coordination tetrahedra to be more flexible.

Table 4. Temperature-related change in the interior angle of metal-organic fences

Material	Temperature (K)	Angle (°)
[Ni(pba)$_2$]•2DMA	112	95.271
[Ni(pba)$_2$]•2DMA	300	94.428
[Co(pba)$_2$]•2DMA	112	96.625
[Co(pba)$_2$]•2DMA	300	95.223
[Zn(pba)$_2$]•2DMA	112	96.342
[Zn(pba)$_2$]•2DMA	300	94.602
[Zn$_{0.77}$Cd$_{0.23}$(pba)$_2$]•2DMA	112	98.809
[Zn$_{0.77}$Cd$_{0.23}$(pba)$_2$]•2DMA	300	95.912
[Cd(pba)$_2$]•2DMA	112	101.844
[Cd(pba)$_2$]•2DMA	300	98.499

Metal-organic framework material [Cd(tib)(dnbpdc) (H$_2$O)]·2DMF·2H$_2$O [tib = 1, 3, 5-tris(1-imidazolyl)benzene, H$_2$dnbpdc = 2, 2′-dinitro-4, 4′-biphenyldicarboxylic acid] was synthesized[25]. The results showed that the compound had a 1-dimensional chain structure joined together by hydrogen bonds so as to generate a 3-dimensional supramolecular structure. The CO_2 and N_2 adsorption behavior was studied, revealing the selective sorption of CO_2.

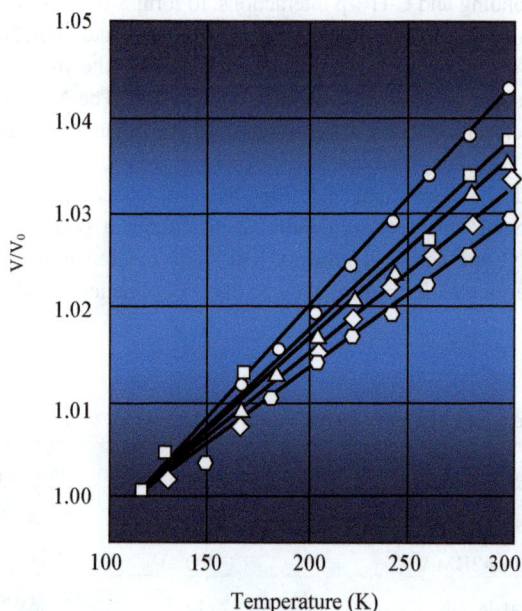

Figure 2. Thermal expansion of [M(pba)$_2$]•2DMA materials
circles: M = Cd; squares: M = Zn,Cd; triangles: M = Zn;
diamonds: M = Co;, hexagons: M = Ni

A pair of supramolecular isomers of Cd[II]-based MOF has been synthesized by utilizing a flexible N,N′-donor linker and a dicarboxylate with excited-state intramolecular proton transfer fluorophore by varying the reaction media[26]. One of the MOFs had a 3-dimensional 4-fold interpenetrating framework with guest solvent in the structure that underwent a solvent-dependent crystalline-to-crystalline structural transformation. The other MOF was structurally rigid in nature and had a 2-fold interpenetrating structure without any guest molecules. Both of the compounds exhibited moderate CO_2 adsorption and one of them, the MOF with the 4-fold interpenetrating structure, also exhibited moderately high H_2 adsorption. Both compounds exhibited an interesting luminescence behavior. In the solid state, the two compounds gave single-peak spectra whereas, upon suspension of the compounds in polar solvents, the maxima split into two peaks with a large Stokes shift. On the other hand, in non-polar solvents, only one emission maximum was observed. This solvatochromic dual-emission phenomenon was due to excited-state intramolecular proton transfer.

Cadmium-based metal–organic framework nanoparticles and a polyurethane–urea elastomer were synthesized[27]. New mixed-matrix membranes were then fabricated from the nanoparticles and the PUU. Scanning electron microscopic images verified that embedding the nanoparticles changes the morphology of the PUU and that the nanoparticles disperse well in the PUU due to a satisfactory compatibility of the polymer and nanoparticles. Fourier transform infra-red spectroscopy and X-ray diffraction analysis confirmed the dispersion of the nanoparticles in the soft segment of the PUU. With increasing temperature, the gas permeabilities of the MMMs improved but their sieving ability deteriorated. An MMM which incorporated 2.5wt% of the MOF showed a CO_2 permeability of ~140 barrer and a CO_2/N_2 selectivity of ~30; 89 and 38% higher than those of the pristine membrane. Gas permeation tests showed that the higher CO_2/N_2 selectivity of the MMMs was due to an improved solubility selectivity while the higher CO_2 permeability was a result of improved CO_2 diffusivity and solubility coefficients.

A novel 2-dimensional metal-organic framework, $[(Cd_2(PDIA)_2 \cdot 3DMF] \cdot 3DMF$ (namely CdPDIA), was constructed[28] by combining an organic ligand 5-(phenyldiazenyl)isophthalic acid (H_2PDIA) and Cd^{II} ion. The resultant sample possesses binuclear $\{Cd_2(COO)_4\}$ clusters and two different modes of the de-protonated organic ligand. The $\{Cd_2(COO)_4\}$ clusters and PDIA2 − ligands can be considered topologically to be 4-connected nodes and 2-connected linkers. Hence, the 2-dimensional structure can be further symbolized as a sql topology with the Schläfli symbol of (44•62). The resultant sample CdPDIA has been characterized and analysed in detail by means of single crystal diffraction, elemental analysis, powder X-ray diffraction and thermogravimetric analysis. Due to the metal sites in the framework of CdPDIA, it can be used as an heterogeneous catalysis for the solvent-free conversion of CO_2 and epoxides into cyclic carbonate. In addition, the resultant sample can be recycled four times without significant activity decrease.

A microporous metal–organic framework, $[Cd(IP)Cl]n$ (HIP = 1H-imidazo[4,5-f][1,10]phenanthroline), with 1-dimensional square channels along the c-axis was constructed[29] using the solvothermal method. Because of the polar nitrogen atoms and the π-electron-rich ligand, activated MOF, the material exhibits a strong interaction with CO_2; with selective capture of CO_2 from a mixture of CO_2/N_2. The MOF also exhibits significant selective capture of CO_2 from CO_2/CH_4 and CO_2/C_2H_4 mixtures due to the sieving-effect of the 1-dimensional channels. The breakthrough experimental results further confirmed that the microporous MOF could be used as a potential porous material for selective CO_2 adsorption.

A method has been developed[30] for adjusting the framework charge of metal-organic materials, and for determining how that charge can affect the band-edge positions and band-gaps of, in particular, those of the anionic Cd^{II} porphyrinic material, $[Cd_{3.2}(H_2TCPP)_2][(CH_3)_2NH_2]_{1.6}$. This is prepared from $H_2TCPP_4^-$, where H_6TCPP is tetrakis(4-carboxyphenyl)-porphyrin and Cd^{II}, thus forming a tubular structure. It has a negatively charged framework, with 60% occupancy of a single type of Cd^{II} ion. An almost neutral form can also be prepared. The $[(CH_3)_2NH_2]^+$ counter-ions can be exchanged with Li^+. Although the surface areas of the material and its Li^+ derivative are only of the order of 407 to $672m^2/g$, CO_2 and CH_4 take-up can attain 44 to 65ml/g and 22 to 26ml/g, respectively, at 1atm and 273K. Those values are nearly tripled at 9atm. The Li^+-exchanged material favoured N_2, CO_2 and CH_4 adsorption; especially at 9atm and 273K. The nature of the counter-ion had little effect upon the band-edge positions and band-gaps, but the framework charge was affected.

Porphyrin metal-organic frameworks are based upon porphyrin or metalloporphyrin ligands and metal nodes and exhibit great thermal and chemical stability and a high CO_2 capture capacity. Because of the easy controllability of the pore size, the catalytic effects can be easily optimized. The 4 types of porphyrin frameworks comprise: low-valence metal ions such as Cu^{2+} and Cd^{2+}, $M_2(COO)_4$ paddle-wheel units with M being Cu^{2+} or Zn^{2+}, infinite metal oxide chains with Al^{3+}, Ga^{3+} or In^{3+} or hard metal oxide clusters with Cr^{3+}, Fe^{3+}, Ti^{4+}, Zr^{4+}, Hf^{4+} … and rare-earth metals. Their main applications are the selective capture of CO_2, CO_2 photoreduction and CO_2 electroreduction. Some catalytic reactions can be carried out at 0.1MPa and room temperature, giving high yields. The CO_2 can be photoreduced to fuels such as methanol and methane by using solar radiation.

It has been noted[31] that sluggish CO_2 mass-transfer in micropores could impair photocatalytic CO_2 reduction. Hybrids of thin porphyrin paddle-wheel framework nanosheets, anchored with gold nanoparticles, have been prepared by electrostatic interaction in order to improve the assembly of gold nanoparticles and porphyrin nanosheets. Thin nanosheets led to faster charge-transfer rates and a higher mass-transport capability, as compared with thick nanosheets, during photocatalysis. On the other hand, plasmonic gold nanoparticles on the surfaces of nanosheets led to more effective light absorption than did nanoparticles which were encapsulated in the matrices of nanosheets. The hybrids exhibited superior photocatalytic CO_2 conversion into HCOOH in an acetonitrile/ethanol system.

Calcium-

The synthesis of SION-8, a novel metal-organic framework based upon Ca^{II} and a tetracarboxylate ligand TBAPy4- endowed with two chemically distinct types of pores characterized by their hydrophobic and hydrophilic properties was described[32]. By altering the activation conditions, two bulk materials were obtained: fully-activated SION-8F and the partially activated SION-8P with the hydrophobic pores alone activated. SION-8P shows a high affinity for both CO_2 (Qst = 28.4kJ/mol) and CH_4 (Qst = 21.4kJ/mol) while, upon full activation, the difference in affinity for CO_2 (Qst = 23.4kJ/mol) and CH_4 (Qst = 16.0kJ/mol) is more pronounced. The intrinsic flexibility of both materials results in a complex adsorption behavior and a greater adsorption of gas molecules than if the materials were rigid. Their CO_2/CH_4 separation performance was tested in fixed-bed breakthrough experiments using binary gas mixtures of various compositions and rationalized in terms of molecular interactions. SION-8F showed a 40-160% increase (depending upon the temperature and the gas mixture composition) of the CO_2/CH_4 dynamic breakthrough selectivity as compared to SION-8P, demonstrating the possibility of modifying the separation performance of a single MOF by manipulating the stepwise activation which was made possible by the biporous nature of the MOF.

A new calcium-based metal-organic framework, $[Ca(NO_2\text{-}BDC)\cdot DMF]_n$, with NO_2-BDC2- = 2-nitroterephthalate, having a 5,5T4 topology and a Schäfli symbol of (48.62)(45.65) was synthesized[33]. It contained 1-dimensional channels, along the crystallographic c-axis, which contained microporosity before activation; with a nanoporous structure obtained upon activation. The 1-dimensional channels were surrounded by $-NO_2$ groups, indicating that this material could be good for CO_2 storage. It exhibited 0.97mmol/g of CO_2 adsorption at 298K at up to 1bar.

Cerium-

A microporous mixed-ligand-based metal-organic framework was synthesized by using two different dicarboxylic acid-based ligands, 4,4'-biphenyldicarboxylate (BPDC) and imino diacetate (IMDA), and two different metal ions, Ce^{3+} and Na^+: $Ce_3Na_3(BPDC)_3(IMDA)_3(DMF)_2(H_2O)_9$. The framework built from Ce-Na-carboxylate layers and BPDC pillars consisted[34] of 2-dimensional slit-shaped pores occupied by extra-framework Na^+ ions. The desolvated framework had a Brunauer–Emmett–Teller surface area of about $771m^2/g$ and a CO_2 take-up of 2.0mmol/g with a CO_2/N_2 selectivity of 68 at room temperature and 1bar. The heat of adsorption was 23kJ/mol, leading to easy CO_2 cycling. The CO_2 positions which were deduced from Monte Carlo simulations showed that Na^+ ions in the channels served as favourable adsorption sites for the oxygen atoms in CO_2 pointing towards the Na^+ ions ($OCO\cdots Na^+$ = 3.34 to 5.87Å), while some

CO_2 molecules sit flat on the phenyl rings of the 4,4′-biphenyldicarboxylate at a CO_2···centroid distance of 3.6 to 3.7Å.

A cerium^{IV}-based metal-organic framework with an alkyne-based linker, acetylenedicarboxylate, was synthesized[35] and found to be made up of octahedral $[Ce_6O_4(OH)_4]^{12+}$ clusters, each of which was connected, to other inorganic units, by 12 acetylenedicarboxylate linkers. This produced a porous network which was analogous to that of UiO-66. The adsorption of CO_2 in this new material was characterised by a high (47kJ/mol) zero-coverage isosteric heat of adsorption. This value was attributed to the presence of the -C≡C- triple-bond in the framework.

The reaction of cerium ammonium nitrate with tetrafluoroterephthalic acid in water leads[36], depending upon the amount of acetic acid used as a crystallization modulator during synthesis, to the appearance of 2 new metal-organic frameworks having UiO-66 and MIL-140 topologies. The F4-UiO-66(Ce) and F4-MIL-140A(Ce) materials both contain pores which are smaller than 8Å, and both contain small amounts of Ce^{III} which preferentially accumulate near to the surface of crystallites. The CO_2 sorption properties of the materials are such that they perform better than do their zirconium-based analogues. The F4-MIL-140A(Ce) has an S-shaped isotherm, with a steep increase in take-up at pressures of less than 0.2bar at 298K, thus making it especially selective for CO_2 as compared with N_2. The calculated selectivity, for a 0.15:0.85 mixture at 1bar and 293K, is greater than 1900 while the calculated isosteric heat of CO_2 adsorption ranges from 38 to 40kJ/mol.

Chromium-

Among the metal-organic framework materials, MIL-101(Cr) has an ultra-high porosity, large window openings and great stability. It comprises trimeric chromium^{III} octahedra which are connected to 1,4-benzenedicarboxylates, and has a large pore size (29 to 34Å) and a specific surface area of more than 3000m^2/g[37]. It was expected to exhibit very high CO_2 capture capacities if functionalized with an alkylamine, but this is not generally observed. An alkylamine was incorporated here[38] by using cyclohexane as a solvent and dispersant. The resultant material exhibited an increased amine loading, and a markedly increased CO_2 capture capacity. The improvement was attributed to a weaker interaction between the alkylamine and solvent, and to a resultant higher chemical potential of the dispersed alkylamine molecules. Tris(2-aminoethyl)amine-modified MIL-101(Cr) exhibited CO_2 take-ups of 4.21mmol/g at 150mbar and 25C, and 4.06mmol/g at 150mbar and 40C. The material also had very different breakthrough times for CO_2 and N_2, and was capable of capturing 4.35mmol/g of CO_2 at 25C or 4.22mmol/g of CO_2 at 40C in a 15:85 gas mixture of CO_2/N_2 (v/v).

A new strategy for CO_2 capture is to create multiple adsorption sites in metal-organic frameworks such as MIL-101(Cr). This can be done[39] by incorporating a diethylenetriamine-based ionic liquid via post-synthesis modification. This liquid, having multi-amine tethered cations and acetate anions could provide additional binding sites and increase the affinity of framework surfaces for CO_2. The high surface area and large cage-size of MIL-101(Cr) meanwhile ensure a better dispersion of the ionic liquid and expose more active sites for CO_2 adsorption. Sufficient free space existed, following functionalization, to facilitate CO_2 transport and permit Cr^{III} sites deep within the pores to be accessed. The additional adsorption sites originating from the ionic liquid and the metal-organic framework synergistically affected the CO_2-capture performance and markedly enhanced the adsorption capacity and selectivity. A high isosteric heat of adsorption reflected the stronger interaction occurring between the composite and CO_2 molecules. In spite of the relatively high initial isosteric heat of adsorption, the composite material could be easily regenerated, with essentially no fall in CO_2 take-up over 6 cycles.

Mixed-matrix membranes were made[40] from an intrinsically microporous polymer, PIM-1, and the crystalline chromium-based terephthalate metal-organic framework material, MIL-101. The latter was in various forms: plain MIL-101, with an average particle size of about $0.2\mu m$, NanoMIL-101, with a particle size of about 50nm, ED-MIL-101, functionalized with ethylene diamine and NH_2-MIL-101, made from 2-aminoterephthalic acid rather than terephthalic acid. The addition of NH_2-MIL-101 and ED-MIL-101 did not increase membrane performance, with regard to CO_2/N_2 and CO_2/CH_4 separation, because of an initial decrease in selectivity at low metal-organic framework content. The O_2 and N_2 permeabilities increased for NH_2-MIL-101. The MIL-101 and NanoMIL-101 caused a marked shift to higher permeability for all gas pairs, and especially CO_2, without any great change in selectivity. The CO_2 permeabilities were among the highest values attained for PIM-1 based mixed-matrix membranes.

The effect of cluster/ligand (X) and cluster/modulator (Y) ratios on the properties of MIL-101/M-X-Y combinations was that the surface area and pore volume of the MIL-101/M-0.5-Y series attained $3596 m^2/g$ and $1.65 cm^3/g$ for MIL-101/M-0.5-0.5. These results[41] were 23.8 and 27.9% higher, respectively, than those for conventional MIL-101(Cr). The CO_2 and H_2S adsorption data at 298K and 1 to 35bar, for MIL-101/M-0.5-0.5, indicated a high adsorption capacity for both CO_2 (3.16mmol/g) and H_2S (7.63mmol/g) at 1bar. These were increases of 44.9 and 59.3%, respectively, over that of conventional MIL-101(Cr). The enhancement in gas adsorption capacity was attributed to the better textural properties of MIL-101/M-0.5-0.5 and to additional unsaturated Cr^{3+} sites which led to stronger interactions at low pressures.

Nanoscale MIL-101(Cr)-Ag/nanoparticle hybrids having sizes ranging from 80 to 800nm were studied[42] under visible light showing that, when the size was reduced to 80nm, the hybrid catalyst exhibited its highest CO_2 photocatalytic reduction activity, with production rates of 808.2μmol/gh for CO and 427.5μmol/gh for CH_4; making it one of the most efficient hybrid catalysts. The high catalytic activity was attributed to the high density of unit cells, on the corners and edges of the catalyst, which were favourable to electron transfer during photocatalytic CO_2 reduction.

The high-porosity MIL-101 was modified by incorporating[43] polyvinylamine via the so-called ship-in-a-bottle technique. The modified material exhibited a much increased CO_2/N_2 selectivity and CO_2 adsorption capacity at low pressure, when a suitable amount of polyvinylamine was present. The adsorption selectivity and capacity were about 11 and 2.5 times those of pristine material, respectively, at 298K. The modified material was easily recyclable and had an isosteric heat of adsorption of 35 to 50kJ/mol over a wide range of levels of CO_2 adsorption. In related work[44], MIL-101(Cr) was modified by introducing polyaniline; again via the ship-in-bottle method. The modified material exhibited a greatly increased adsorption capacity for CO_2, especially at 0.15atm, even though the original porosity was decreased by modification. The modified material also exhibited an increased selectivity for CO_2 over N_2. The improved adsorption of CO_2 was attributed to basic species on the well-dispersed polyaniline.

A new and efficient multifunctional catalytic system for the cyclo-addition of carbon dioxide with epoxides to synthesize cyclic carbonates under mild and solvent-free reaction conditions has been developed. The catalytic tests revealed[45] that [P12,4,4,4]Br/MIL-53(Cr) was the best and most powerful catalytic system in cyclo-addition with excellent yields (96–99%) under solvent-free condition and 100C, 1.0MPa for 2 to 3h. The synergistic effect of anion and cation of ionic liquid [P12,4,4,4]Br as well as the chromium site of co-catalyst, MIL-53(Cr), contributed to the excellent catalytic activity.

The effect of metal oxide loading on the CO_2 absorption by chromium[III] chloride hexahydrate and trimesic acid was investigated[46]. The fabricated MOF were characterized using scanning electron microscopy, with energy dispersive X-ray spectroscopy to observe its surface morphology and composition. The SEM morphology suggested that Cr-TA-3 possesses the highest surface area to volume ratio; attributed to the formation of an agglomerate of crystal rods structure. Carbon dioxide gas adsorption was measured by thermogravimetric analysis, indicating that Cr-TA-3 exhibited the highest CO_2 adsorption: 0.038mmol/g.

A composite which was based upon MIL-101 $(Cr_3F(H_2O)_2O(BDC)_3$, where BDC is terephthalic acid, loaded with N,N'-dimethylethylenediamine, exhibited[47] superior carbon dioxide capture even though the Brunauer–Emmett–Teller specific surface area had decreased by 52% as compared with that of MIL-101. At 273 and 298K, under 1bar, dimethylethylenediamine-loaded samples exhibited 8.7 and 18.7% increases, respectively, in CO_2 capacity: from 2.30 to 2.50mmol/g at 273K, and from 1.35 to 1.59mmol/g at 298K. The improved capacity and selectivity were attributed to the higher (33.9kJ/mol) isosteric heat of CO_2 adsorption at zero coverage, which was in turn due to interactions between amines and CO_2.

The separation of CO_2 and N_2O mixtures is still a challenge due to their identical molecular masses and sizes. On the other hand, CO_2 is acidic whereas N_2O is not, and so it is feasible that they could be separated on this basis. The use of an ethylenediamine–functionalized metal organic framework ED-MIL-100Cr-0.2/0.4/0.6 in the construction of a new alkaline adsorption site for CO_2 binding was considered. The experimental results showed[48] that ED-MIL-100Cr-0.4 with the highest adsorption heat of CO_2 of up to ~80kJ/mol with that of N_2O being 25kJ/mol, has the highest equilibrium adsorption selectivity (28.0, IAST method) of CO_2 versus N_2O so far reported. It was also confirmed that ED-MIL-100Cr-0.4 offered the best performance with regard to the separation of CO_2 and N_2O mixtures in breakthrough tests.

Cobalt-

Cobalt 2-methylimidazolate frameworks (ZIF-67) have been synthesized[49] by using solvothermal methods, yielding samples with a space-group of $I\overline{4}3m$ and a cell-size of 17.0545Å. The cobalt was tetrahedrally coordinated to the nitrogen atoms of the 2-methylimidazole linker, rendering the metal sites saturated and largely irrelevant to gas adsorption. The gas-adsorption properties of ZIF-67 with regard to H_2 and CO_2 guest molecules at 40bar indicated storage capacities of 0.28wt% at 300K and 2.84wt% at 77K. At 300K under 1bar, the material adsorbed 3.9wt% of CO_2 but, at a pressure of 50bar, the CO_2 adsorption capacity was 51.4wt%.

A porous 3-dimensional Co^{II} metal-organic framework, $[\{Co(TCPB)\ 0.5(H_2O)\}\cdot DMF]_n$ (MOF-1), (where H_4TCPB = 1,2,4,5-tetrakis(4-carboxyphenyl)benzene) constituting a rigid tetratopic H_4TCPB ligand was synthesized[50] under solvothermal conditions. Single-crystal X-ray structure determination revealed the 3-dimensional framework structure of MOF1 with 1-dimensional channels having sizes of 10.5Å x 10.5Å decorated with Co^{II} ions on the crystallographic a-axis. Gas-adsorption studies showed the selective adsorption property of MOF1 for CO_2 over other $(N_2, Ar$ and $H_2)$ gases with an isosteric heat of adsorption of 35.4kJ/mol. This high value was attributed to the stronger interaction of CO_2 molecules with the Co^{II} metal sites lining the 1-dimensional channels of MOF1. Interestingly, the coordinated water molecules at the Co^{II} centers can be removed by activating MOF1 at a temperature of 120C to generate a framework with pores lined with unsaturated Lewis acidic Co^{II} ions. The activated MOF1 shows a very good catalytic activity for the highly selective, solvent-free, heterogeneous conversion of CO_2 into cyclic carbonates under mild conditions involving an atmospheric pressure of CO_2. Furthermore, the MOF1 catalyst was easy to recycle and re-use over several cycles with no substantial loss of catalytic activity.

Flexible tetracarboxylic acids having ether (H4AOIA) or amine (H4ANIA) linkages have been used to form 2 distinct cobalt-based metal-organic frameworks[51]. Both exhibited a marked ability to catalyse the chemical conversion of CO_2 into cyclic carbonates under ambient reaction conditions. Carbonization of the 2-dimensional Co-ANIA structure, under inert atmospheric conditions at 800C, resulted in the formation of a cobalt-containing N-doped nanocomposite which acted as an efficient non-noble metal electrocatalyst.

The cobalt-based metal-organic framework, $[Co_3(L)(OH)_2(H_2O)_4]\cdot 2DMF\cdot 2H_2O$, was synthesized[52] under solvothermal conditions using pyridyl-decorated tetracarboxylic acid, 2,6-di(2′,5′-dicarboxylphenyl)pyridine (H4L). Structural analysis demonstrated that it was a 3-dimensional framework based upon 1-dimensional alternating Co_4 chain units. The de-solvated structure contains 1-dimensional open channels with a highly polar pore surface decorated with open metal sites, μ_3-OH group and pyridyl group sites, exhibiting multipoint interactions between C_2H_2 and CO_2 molecules. The framework efficiently takes up C_2H_2 and CO_2, with a significant selectivity for C_2H_2 and CO_2 over CH_4.

A microporous metal-organic framework, $\{[Co_2(4,4′\text{-bpy})(L)]\cdot H_2O\cdot 0.5(DMF)\}_n$, was obtained from the self-assembly of cobalt^{II} nitrate hexahydrate, rigid tetrapodal carboxylic acid 4,4′,4″,4‴-silanetetrayltetrabenzoic acid (H4L) and a rigid bifunctional linker, 4′4′-bipyridine (4,4′-bpy), under solvothermal conditions[53]. It was characterized by Fourier transform infra-red spectroscopy, thermogravimetric analysis, elemental analysis, and powder and single crystal X-ray diffraction. Its single-crystal structure reveals the presence of a Co^{II}-paddle-wheel core as the SBU, which is extended to form a doubly interpenetrated 3-dimensional framework with a (4,6)-connected sqc422-type uninodal net topology with the Schläfli point symbol $\{4\,2\cdot 5\,10\cdot 7\,2\cdot 8\}\{4\,2\cdot 5\,4\}$. There are two types of open channels in this framework (rhombic and trigonal), which run along all three axes. Its thermal and chemical stabilities were established based upon thermogravimetric analysis and *in situ* variable temperature powder X-ray diffraction. The activated framework (lattice solvent free) of the compound exhibits a modest take-up of CO_2 (53.8 and 36.4cm^3/g at 273 and 298K at 1bar pressure, respectively), with moderately high selectivities for CO_2/N_2 and CO_2/CH_4 gas separation under ambient conditions (298 and 273K under 1bar of pressure).

The metal–organic frameworks, $M(BPZNO_2)$ (M = Co, Cu, Zn; H_2BPZNO_2 = 3-nitro-4,4′-bipyrazole), were prepared via solvothermal routes[54]. They showed good thermal stability both under a N_2 atmosphere and in air, with decomposition temperatures peaking up to 663K for $Zn(BPZNO_2)$. Their crystal structure is characterized by 3-dimensional networks with square (M = Co, Zn) or rhombic (M = Cu) channels decorated with polar NO_2 groups. As revealed by N_2 adsorption at 77K, they are micro-mesoporous materials with Brunauer–Emmett–Teller specific surface areas ranging from 400 to $900m^2/g$. Remarkably, under the mild conditions of 298K and 1.2bar, $Zn(BPZNO_2)$ adsorbs 21.8wt%CO_2 (4.95mmol/g). It exhibits a Henry CO_2/N_2 selectivity of 15 and an ideal adsorbed solution theory (IAST) selectivity of 12 at 1bar. As a CO_2 adsorbent, this compound is the best-performing MOF to date among those bearing a nitro group as a unique chemical tag. High-resolution powder X-ray diffraction at 298K and various CO_2 loadings revealed, for the first time in a NO_2-functionalized MOF, the insurgence of primary host–guest interactions involving the $C(3)–NO_2$ moiety of the framework and the oxygen atoms of carbon dioxide, as confirmed by Monte Carlo simulations. This interaction mode is markedly different to that observed in NH_2-functionalized MOFs, where the carbon atom of CO_2 is involved.

A new polyhedron-based metal-organic framework, namely $\{[Co_6(OH)_2(H_2O)_4(cpt)_9](NO_3)(DMF)_{13}\}$ (Hcpt = 4-(4′-carboxyphenyl)-1,2,4-triazole), was constructed[55] by employing a bifunctional triazolyl-carboxyl ligand Hcpt. Single-crystal X-ray structural analysis showed that the compound features a rare hexanuclear $\{Co_6(OH)_2(H_2O)_6\}10+$ cluster and could be viewed topologically as being a 10-connected bct net. Furthermore, the compound comprises octahedral cages with an inner diameter of 19.6Å x 12.9Å, and 2-dimensional pore systems along the a- and b-axes with a high density of open metal centers generated by the removal of coordinated water molecules, which contributes to a high CO_2 adsorption capacity and significantly selective capture of CO_2 over CH_4 at around room temperature. In addition, the resultant activated compound could behave as an heterogeneous Lewis catalyst and facilitate the chemical fixation of CO_2 coupling with epoxides to give cyclic carbonates.

A flexible metal-organic framework, $[Co_2(OBA)_2(BPMP)]_n$ (COB), with a new network topology was reported[56]. COB displays structural flexibility under CO_2 gas pressure at 298K, and the resultant porous phases have been characterized by in situ X-ray diffraction analysis. Activation yielded a framework with discrete voids and a substantial reduction in guest-accessible volume. Single-crystal X-ray diffraction analysis under controlled CO_2 pressure shows that COB exhibits a breathing mode of flexibility, combined with an overall swelling of the framework. This combination of mechanisms is highly unusual.

Three Co^{II} metal-organic frameworks, namely, $\{[Co_2(L)_2(OBA)_2(H_2O)_4]_xG\}_n$, $\{[Co(L)_{0.5}(OBA)]_xG\}_n$ and $\{[Co_2(L)_2(OBA)_2 (H_2O)]\cdot DMA\cdot xG\}_n$ [where L = 2,5-bis(3-pyridyl)-3,4-diaza-2,4-hexadiene, H_2OBA = 4,4′-oxybisbenzoic acid, DMF = dimethylformamide, DMA = dimethylacetamide and G denotes disordered guest molecules], have been synthesized[57] under diverse reaction conditions via the self-assembly of a bent dicarboxylate and a linear spacer with a Co^{II} ion. While $\{[Co_2(L)_2(OBA)_2(H_2O)_4]_xG\}_n$ crystallized at room temperature in DMF to form a 2-dimensional layer structure, $\{[Co(L)_{0.5}(OBA)]_xG\}_n$ is formed by the assembly of similar components under solvothermal conditions with a 3-dimensional network structure. On the other hand, by changing the solvent to DMA, $\{[Co_2(L)_2(OBA)_2 (H_2O)]\cdot DMA_xG\}_n$ could be crystallized at room temperature with a 3-dimensional architecture. Of the three compounds, activated $\{[Co(L)_{0.5}(OBA)]_xG\}_n$ was found to be permanently microporous in nature, with a Brunauer–Emmett–Teller surface area of $385m^2/g$, and exhibited a moderately high take-up capacity for C_2H_2 and CO_2 while taking up much less CH_4 and N_2 under ambient conditions. As a result, high ideal adsorbed solution theory separation selectivities are obtained for CO_2/N_2 (15:85), CO_2/CH_4 (50:50), and C_2H_2/CH_4 (50:50) gas mixtures, making $\{[Co(L)_{0.5}(OBA)]_xG\}_n$ a potential candidate for these important gas-separation tasks under ambient conditions.

A 3-fold interpenetrated metal-organic framework $[Co_2(OBA)_4$ $(PTD)\cdot 3DMF\cdot CH_3CH_2OH\cdot 5H_2O]_n$ has been synthesized by utilizing 4,4′-oxybis(benzoic acid) (H_2OBA) as the linker, 6-(pyridin-4-yl)-1,3,5-triazine-2,4-diamine (PTD) as the ligand, and $CoCl_2\bullet 6H_2O$ via a solvothermal method[58]. The compound exhibits not only a high take-up capacity for CO_2 molecules with an estimated high sorption heat (50.6kJ/mol at zero loading), but also a significant selective adsorption of CO_2 over CH_4, which can be attributed to the presence of proper-sized pores with high polarity, amine groups and triazine rings of PTD linker decorating the pores. Monte Carlo simulations of CO_2 adsorption of the compound demonstrated that CO_2 molecules are preferentially adsorbed around the PTD ligands. Furthermore, the complex displays a relatively high adsorption capacity of H_2 ($101.7cm^3/g$ at 1bar) at 77K.

Equilibrium adsorption experiments were used to demonstrate that the flexible metal-organic framework Co(bdp) (bdp2- = 1,4-benzenedipyrazolate) exhibits a large CO_2 adsorption capacity and approaches complete exclusion of CH_4 for 50:50 mixtures of the two gases, leading to outstanding CO_2/CH_4 selectivity under these conditions. *In situ* powder X-ray diffraction data indicate[59] that this selectivity arises from reversible guest templating, in which the framework expands to form a CO_2 clathrate and then collapses to the non-templated phase upon desorption. In an atmosphere dominated by CH_4, Co(bdp) adsorbs minor amounts of CH_4 together with CO_2, highlighting the importance of studying all relevant pressure and composition ranges via multicomponent measurements when examining mixed-gas selectivity in structurally flexible materials. Altogether, these results show that Co(bdp) may be a promising CO_2/CH_4 separation material and provide insights into the further study of flexible adsorbents for gas separation.

Two 3-dimensional Co^{II}-MOFs with open metal sites and exposed azo functionality on the MOF backbone have been constructed via mixed-ligand assembly[60]. Both of the MOFs, (1) $\{[Co_2(1,4\text{-}NDC)_2(L)(H_2O)_2(\mu_2\text{-}H_2O)]\cdot(DMF)_2(H_2O)\}_n$ and (2) $\{[Co(fma)(L)(H_2O)_2]\cdot S\}_n$ [1,4-NDC = 1,4-naphthalene dicarboxylic acid, fma = fumaric acid, L = 3,3′-azobis pyridine and S = disordered solvents] exhibit 3-dimensional frameworks with metal-bound aqua ligands. These metal-bound aqua ligands, as well as the lattice solvent molecules, could simply be removed upon activation; thus giving desolvated frameworks 1a and 2a, respectively, while maintaining the original crystallinity with a varying number of open metal sites. Although crystallographic analysis revealed a porous structure for both of the MOFs (34.4% and 14.3% void volume for 1 and 2, respectively), 1 showed a permanently microporous nature with a Brunauer-Emmett-Teller surface area of $197 m^2/g$ and moderate CO_2 take-up capacity as established through a gas sorption study. Both MOFs exhibit efficient catalytic activity for the chemical fixation of CO_2 to give cyclic carbonates in the presence of a co-catalyst, cyanosilylation reaction, and Knoevenagel condensation under solvent-free and mild conditions; thus demonstrating their multipurpose heterogeneous catalytic nature. The limited pore space, decorated with exposed metal sites, and the functional azo-groups were efficient for size-selective heterogeneous catalysis with varying catalytic efficiencies.

The Co^{II} metal-organic framework, $\{[Co(\mu_3\text{-}L)(H_2O)]\cdot0.5H_2O\}_n$ (H_2L = thiazolidine 2,4-dicarboxylic acid) with rich Lewis acid sites was used[61] as a catalyst for the conversion of CO_2 and propylene oxide into propylene carbonate with a yield of up to 98% at 50C and 1atm. The compound exhibited excellent reusability, and could be regenerated easily during at least 5 runs without any decrease in the yield. Two types of nano-crystals (N1 and N2) of the compound with polyvinylpyrrolidone and hexadecyltrimethylammonium bromide as surfactants, respectively, were prepared and their catalytic properties were compared with those of the compound in the powder phase. A significant improvement in catalytic efficiency and product yield was observed when the compound was nano-crystallized.

Although the selective separation of CO_2 from gas mixtures has attracted great attention and although porous MOF materials are a promising means for achieving such separation, their stability in the presence of moisture must be taken into account. A microporous metal–organic framework, $\{[Co(OBA)(L)_{0.5}]\cdot S\}_n$ (IITKGP-8), was constructed[62] by employing a V-shaped organic linker with an azo-functionalized N,N′ spacer forming a 3-dimensional network with mab topology and 1-dimensional rhombus-shaped channels along the crystallographic b-axis with a void volume of 34.2 %. The activated MOF reveals a moderate CO_2 take-up capacity of 55.4 and $26.5 cm^3/g$ at 273 and 295K/1bar, respectively, whereas it takes up a significantly lower amount of CH_4 and N_2 under similar conditions and thus exhibits its potential for the highly selective sorption of CO_2, with an excellent IAST selectivity of CO_2/N_2 (106 at 273K and 43.7 at 295K) and CO_2/CH_4 (17.7 at 273K and 17.1 at 295K) under 1bar. More importantly, this MOF exhibits an excellent moisture stability as assessed by PXRD experiments coupled with surface-area analysis.

A new microporous metal-organic framework with the formula, $\{Co_2(oba)4(3\text{-}bpdh)_2\}4H_2O$ [oba = 4,4′-oxybis(benzoic acid); 3-bpdh = N,N′-bis-(1-pyridine-3-yl-ethylidene)-hydrazine], was assembled and its morphology was found[63] to undergo a microrod-to-nanosphere transformation with temperature change. Core-shell Au-Pd functional nanoparticles were successfully encapsulated within the center of the monodisperse nanospheres, and platinum nanoparticles were well-dispersed and fully immobilized on the surface of Au-Pd-1Co to build Pt/Au-Pd-1Co composites which exhibited nanoparticle catalytic activity for the reverse water-gas shift-reaction. The core-shell Au-Pd nanoparticles in metal-organic frameworks markedly increased the CO selectivity of the catalyst, and the platinum nanoparticle loading on the surface of the nanosphere led to CO_2 conversion.

Two dual-linker metal-organic frameworks were prepared[64] with metal salts, terephthalic acid and 4, 4′-bipyridine as precursors, and subsequently used for preparing cyclic carbonates from carbon dioxide and epoxides. These two MOFs were characterized by a number of techniques. They showed superior catalytic performance under solvent-less and co-catalyst-free conditions: Co(tp)(bpy) achieved epichlorohydrin conversions as high as 95.75% and chloropropylene carbonate yields of 94.18% under optimum reaction conditions. The high activity of Co(tp)(bpy) was attributed to the coexistence of Lewis acidic and basic active sites on the catalyst, derived from incompletely coordinated metal cations and uncoordinated pyridine groups, respectively. In addition, Co(tp)(bpy) maintained this high catalytic performance after five consecutive reaction cycles.

Highly porous polyhedral metal-organic frameworks of Co^{II}/Ni^{II}, $\{[M_6(TATAB)_4 (DABCO)_3 (H_2O)_3]\cdot 12DMF\cdot 9H_2O\}_n$ (where M = Co^{II} (1)/Ni^{II} (2), H_3TATAB = 4,4′,4″-S-triazine-1,3,5-triyl-tri-P-aminobenzoic acid, and DABCO = 1,4-diazabicyclo[2.2.2]octane) have been synthesized[65] solvothermally. Both of these MOFs had a 2-fold interpenetrated 3-dimensional framework structure composed of dual-walled cages of about 30Å, functionalized with a high density of Lewis acidic Co^{II}/Ni^{II} metal sites and basic-NH-groups. Interestingly, the former MOF exhibited the selective adsorption of CO_2, with a heat of adsorption of 39.7kJ/mol; further supported by theoretical studies giving a computed binding energy of 41.17kJ/mol. The presence of a high density of both Lewis acidic and basic sites made both MOFs ideal candidates for achieving the cocatalyst-free cyclo-addition of CO_2 to epoxides. The MOFs could also act as excellent recyclable catalysts for the cyclo-addition of CO_2 to epoxides in the high-yield synthesis of cyclic carbonates under co-catalyst-free mild conditions of 1bar of CO_2. The former MOF could be recycled 5 times successively with no substantial loss of catalytic activity.

The concept of the ultramicroporous building unit permits the creation of hierarchical bi-porous features that act together so as to increase gas take-up capacity and separation ability. The smaller pores promote selectivity, while the larger inter-unit packing pores increase the take-up capacity. In a cobalt-based metal-organic framework material having octahedral cages with a 4.5Å pore size, the C_2H_2 take-up capacity at 1atm attained 193.0cm^3/g (8.6mmol/g) at 273K and 134.0cm^3/g (6.0mmol/g) at 298K. The C_2H_2/CO_2 selectivity was up to 8.5 at 298K. Dynamic breakthrough studies at room temperature and 1atm indicated a C_2H_2/CO_2 breakthrough time of up to 79min/g. Monte Carlo simulations[66] attributed the ultra-high C_2H_2/CO_2 selectivity mainly to the ultramicropores, while the packing pores promoted C_2H_2 take-up capacity.

The isoreticular expansion and functionalization of charged-polarized porosity was investigated[67] by using 11 isostructural zwitterionic metal-organic frameworks of the form, [$M_3F(L1)_3(L2)_{1.5}$], which were prepared via the solvothermal reaction of cobalt and nickel tetrafluoroborates with binary ligands composed of zwitterionic pyridinium derivatives and functionalized ditopic carboxylate auxiliary ligands: 1-(4-carboxyphenyl)-4,4′-bipyridinium chloride, 1-(4-carboxyphenyl-3-hydroxyphenyl)-4,4′-bipyridinium chloride, benzene-1,4-dicarboxylic acid, 2-aminobenzene-1,4-dicarboxylic acid, 2,5-dihydroxy-1,4-benzenedicarboxylic acid, biphenyl-4,4′-dicarboxylic acid, or stilbene-4,4′-dicarboxylic acid. The materials had cubic symmetry (I$\bar{4}$3m, a = 31-36Å) with a 3-dimensional pore system and a void space of 73 to 81%. The pore system comprised 3 types of pore, which could be varied in size from 17.4 to 18.8Å, from 8.2 to 12.8Å and from 4.8 to 10.4Å, depending upon the auxiliary ligand. The materials had non-interpenetrating structures, with permanent porosity, and were stable at up to 300C. The Brunauer–Emmett–Teller surface areas ranged from 1250 to 2250m^2/g.

Microporous Co-ABTC, where H_4ABTC is 3,3',5,5'-azobenzene tetracarboxylic acid, as synthesized by solvothermal reaction[68], contained spherical pores with a radius of 5.2Å. The Brunauer–Emmett–Teller area was $585.9m^2/g$, and the activated material exhibited a C_2H_2/CO_2 selectivity of 4.28 at 298K and extremely low pressures. Microporous Cu-ABTC, $Cu_6(ABTC)_3(H_2O)_6\bullet(DMA)_7(H_2O)_{12}$, was prepared[69] in the same way. The activated material had a Brunauer–Emmett–Teller specific surface area of $2366m^2/g$. Due to the open Cu^{2+} functional sites, the material had a C_2H_2 storage capacity of $208cm^3/g$ at 298K and good separation selectivities for equimolar C_2H_2/CH_4 and CO_2/CH_4 at 298K.

The synthesis and structure were reported[70] for Co-MOF-1, $[CO_2(dpmndi)(bdc)_2)]\cdot DMF$; (dpmndi = N,N'-bis(4-pyridylmethyl)-1,4,5,8-naphthalene di-imide and bdc = benzene-1,4-dicarboxylate), together with an isonicotinate based Co-MOF-2, $[Co_3(ina)_4(OH)(H_2O)_3]\cdot H_2O\cdot NO_3$. Single-crystal X-ray analysis revealed that Co-MOFs 1 and 2 had a (5)-connected and (8)-connected 3-dimensional framework architecture with tts and hex topology, respectively. Importantly, Co-MOF-1 contained a microsized cage with an internal cavity of 8.6Å. Gas-sorption analysis revealed a high take-up of N_2 and CO_2, as compared with that of H_2 and CH_4. Both Co-MOFs had a high value of isosteric heat of adsorption for CO_2 gas.

The synthesis and characterization of a mixed ligand metal-organic framework with good thermal and chemical stability, $\{[Co(BDC)(L)\bullet 2H_2O]_xG\}_n$ (CoMOF-2), involving an aromatic dicarboxylate (H_2BDC = 1,4-benzenedicarboxylic acid) and an acyl-decorated N-donor linker [L = (E)-N'-(pyridin-4-ylmethylene) isonicotinohydrazide] by various physicochemical techniques, including single-crystal X-ray diffraction, are reported[71]. The metal-organic framework showed a good affinity for CO_2 capture, and Monte Carlo simulation studies exposed strong interactions of CO_2 with the functionalized N-donor ligand of the framework. CoMOF-2 and KI acted as an efficient binary catalyst for the sustainable utilization of CO_2 with spiro-epoxy oxindole to spirocyclic carbonate under ambient conditions. MOF-based catalysis was reported for the cyclo-addition of oxindole-based epoxides with CO_2 for harvesting new spirocyclic carbonates.

Three Co-based isostructural MOF-74-III materials with expanded pores were synthesized, with various numbers of fused benzene rings on the side-chains of same-length ligands, so as to tailor the pore sizes to 2.6, 2.4 and 2.2nm[72]. The gas sorption results for these highly mesoporous materials showed that alternately arranged fused benzene rings on one side of the ligand could serve as extra anchoring sites for CO_2 molecules with π–π interactions; thus markedly enhancing CO_2 take-up and CO_2/CH_4 and CO_2/N_2 selectivity. Greater steric hindrance with regard to open Co^{II} sites was imposed by ligands flanked with fused benzene rings on both sides; compromising extra-site enhancement. In the catalytic conversion of CO_2 with propylene oxide, to form propylene carbonate, as-synthesized MOF-74-III(Co) with highly exposed and accessible open Co^{II} centers, large mesopore apertures and multi-interactive sites led to a higher catalytic activity as compared with that of other metal-organic frameworks. Benzene rings which were fused to ligands hampered the functionality of Co^{II} centers as Lewis acid sites.

The support of metal–organic frameworks on substrates such as hollow fibers is a practical route to large-scale industrial application. An effective approach is the growth of framework materials on the surface of hollow carbon fibers produced by the pyrolysis of cross-linked hollow Torlon fibers[73]. The framework/carbon composites, based upon hollow carbon fibers were functionalized in different media so as to increase the number of surface hydroxyl groups before metal-organic framework growth. The latter were incorporated by growing MOF-74 and UTSA-16 within the pores, and on the outer surfaces, of hollow fibers by means of dip-coating and layer-by-layer methods. Composites having relatively high framework loadings of 37 to 38%, and porous structures, were prepared which had film thicknesses of 10 to 15μm. The MOF-74/carbon and UTSA-16/carbon composites had surface areas of 266 and 211m^2/g, and pore volumes of 0.28 and 0.20cm^3/g, respectively. The composites had CO_2 adsorption capacities of 2.0 and 1.2mmol/g in the case of UTSA-16 and MOF-74, respectively, at room temperature under 1bar.

Switchable pillared layer metal-organic frameworks, $M_2(2,6\text{-ndc})_2(\text{dabco})$ (DUT-8(M), M = Ni, Co, 2,6-ndc = 2,6-naphthalenedicarboxylate, dabco = 1,4-diazabicyclo-[2.2.2]octane, were synthesised in two different crystallite size regimes to produce particles of up to 300μm and smaller particles of around 0.1μm. The textural properties and adsorption-induced switchability of the materials, obtained from both syntheses, were studied via the physisorption of N_2 at 77K, CO_2 at 195K and n-butane at 273K, revealing[74] pronounced differences in adsorption behavior for Ni and Co analogues. While the smaller nano-sized particles (50 to 200nm) were rigid and showed no gating transitions, thus confirming the importance of crystallite size, the large particles exhibited a pronounced switchability, with characteristic differences for the two metals resulting in distinct recognition effects for various gases and vapours. The adsorption of various vapours demonstrated a higher energetic barrier for the gate-opening of DUT-8(Co) as opposed to DUT-8(Ni), because the gate-opening pressure of the cobalt-based material was shifted to a higher value for the adsorption of dichloromethane at 298K. There were clear geometrical differences in the paddle-wheel units of the metal-organic frameworks.

Table 5. Properties of Cu–BTC, including the CO_2 adsorption at 1bar

Sample	S_{BET} (m^2/g)	Take-Up$_{0C}$ (mmol/g)	Take-Up$_{25C}$ (mmol/g)	Cu/BTC
Cu–BTC	1415	8.59	4.74	1.81
Cu–BTC(80%)	1506	9.06	5.05	1.45
Cu–BTC(60%)	1560	9.33	5.15	1.09
Cu–BTC(40%)	1572	9.07	5.08	0.72

Copper-

Copper–BTC (HKUST-1) is one of the most widely studied metal-organic frameworks. The material is easily assembled from copper cations and benzene-1,3,5-tricarboxylate (BTC) ligands to give a microporous face-centred cubic network. During synthesis, a fraction of the copper can be advantageously replaced by magnesium in order to increase CO_2 adsorption, as reflected by a higher specific Brunauer–Emmett–Teller (BET) surface area, larger micropore volume and a CO_2 adsorption of 9.33mmol/g at 0C and 1bar. Copper–BTC was prepared here[75] with a metal–ligand ratio of 1.81:1.0 in a 1:1 mixture of ethanol and water, with the molar ratio of magnesium arranged to be 20, 40 or 60% of the total number of metal ions. The incorporation of magnesium was carried out by varying the amount of magnesium salt during synthesis. Samples which were prepared by adding 40% of magnesium exhibited the greatest CO_2 adsorption: 9.31 and 5.15mmol/g at 0 and 25C (1bar), respectively. Samples having a metal–ligand ratio of 1.09:1.0 had a specific BET surface area of 1560m^2/g and a micropore volume of 0.70cm^3/g. By controlling the metal–ligand ratio, rather than the magnesium addition, it was possible to adjust the specific BET surface area, micropore volume and CO_2 adsorption capacity. The CO_2 adsorption performance was determined at 0 and 25C (table 5). At both 0 and 25C, the CO_2 adsorption exhibited a similar trend to the changes observed in BET surface area and microporosity. An initial increase in the CO_2 take-up with decreasing fraction of copper attained a maximum and then decreased. An increase in CO_2 adsorption with increasing pressure was also observed (figure 3). It was assumed that side-pockets were occupied first, followed by unoccupied positions around unsaturated copper atoms and, finally, square channels. Samples having a 1.09:1.0 metal–ligand ratio had the highest specific BET surface area (1560m^2/g), a micropore volume of 0.70cm^3/g and a micropore size of 0.55nm, leading to the highest CO_2 adsorption (9.33mmol/g) at 0C and 1bar. Samples of Cu–BTC(60%) exhibited a high CO_2/N_2 selectivity and good recycling ability.

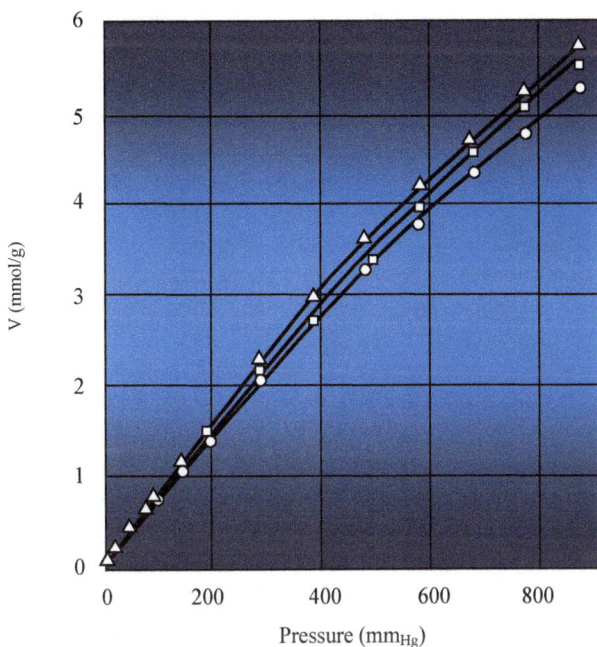

Figure 3. CO$_2$ adsorption isotherms of Cu–BTC compositions at 25C triangles: Cu–BTC (60%), squares: Cu–BTC (80%), circles: Cu–BTC

Mixed-matrix membranes have been prepared[76] which comprised metal-organic frameworks of amine-modified Cu-BTC, NH$_2$-Cu-BTC, and sub-micron amine-modified Cu-BTC, sub-NH$_2$-Cu-BTC; all incorporated into Pebax-1657 polymer. There was an increase in the surface roughness, and in the presence of amino-groups at the surface, following amination. There was a decrease in the size of framework crystals following sub-micron amination. Both of the modified materials exhibited greater CO$_2$ adsorption than that of unmodified Cu-BTC. The modified materials had a greater compatibility with the polyether-block-amide matrix in mixed-matrix membranes. The Pebax/sub-NH$_2$-Cu-BTC membrane enjoyed an improved CO$_2$/N$_2$ and CO$_2$/CH$_4$ selectivity, at the cost of a slight CO$_2$ permeability.

It is still necessary to take account of the effect of processing conditions on a given material, and a study has been made[77] of the influence of pressure and time upon the chemical structure and crystallinity of Cu-BTC and MIL-53(Al) metal-organic framework tablets. A no-binder method was used, with pressures of 3.7, 7.4, 29.6 or 59.2kN/m^2 being applied for 30, 60 or 120s. The resultant tablets were crushed, and sieved to give fractions ranging from 500 to 650μm. Thermogravimetry was then used to assess the effect of the shaping of metal-organic frameworks upon their CO_2 adsorption, and this revealed that compression has a marked effect upon the crystal structure and the CO_2 adsorption ability. A soft-template-assisted method for the preparation of Cu-BTC has been proposed[78] in which the supramolecular assembly of di-cationic quaternary ammonium structure-directing agents interacts cooperatively with Cu-BTC framework coordination layers via weak electrostatic interactions and self-assemble to form Cu-BTC which contains a hierarchical system of mesopores interconnected by micropores. The resultant metal-organic framework has a bimodal pore-size distribution of the micropores and mesopores, and samples which were prepared using a di-cationic structure-directing agent and swelling agent exhibited a further improvement in mesopore diameter. The hierarchical framework had a greater CO_2 take-up capacity than that of conventional Cu-BTC. Cu-BTC coatings on carbon-fiber paper have been prepared[79], by using seed-induced hydrothermal methods, in order to improve the H_2 adsorption, CO_2/H_2 selectivity and thermal/electrical conductivities of the pure Cu-BTC. The H_2 adsorption capacity of coatings treated for 9h was similar to that of pure Cu-BTC. The selectivity for equimolar CO_2/H_2 over 18h was 58 to 209 at 0 to 100kPa and 273.15K. The effective thermal conductivity of the composite was 17 to 25 times higher than that of pure Cu-BTC, while the electrical conductivity was 1.3 x 10^{11} to 1.6 x 10^{11} times greater than that of pure Cu-BTC.

Metal-organic frameworks based upon copper (Cu-BTC) and zirconium (UiO-66) were prepared[80] by using solvothermal methods, and the effects of temperature, reaction time and CO_2 pressure were studied for a standard set of conditions: 0.16mol% of catalyst, 60C, 8h and 1.2MPa CO_2 pressure. The Cu/Zr frameworks were used for CO_2-epoxide cyclo-addition reactions, with a tetrabutylammonium bromide co-catalyst. The UiO-66/Cu-BTC combination led to the efficient conversion of epichlorohydrin, with better than 99% selectivity. The extensive conversion of the epichlorohydrin was attributed to the synergistic effect of the copper and zirconium, and to the bromine from tetrabutylammonium bromide.

Three BTC-based frameworks which differed in morphology were characterized[81] by means of breakthrough experiments in order to determine the effect of the structural properties on the CO_2 absorption capacity. The chosen materials were Zn-HKUST-1 because of its low surface area and coordinate-wise unsaturated metal sites, Al-MIL-96 because of the environment within its pores and Fe-MIL-100 because of its microporous nature and high surface area. The order of CO_2 take-up was Al-MIL-96 > Zn-HKUST-1 > Fe-MIL-100. In every case, the chemistry of the pores proved to have a greater effect upon CO_2 absorption capacity than did the porosity itself under post-combustion conditions.

Composites for CO_2 capture were prepared[82] by combining the Cu-BTC framework with porous carbon-based materials such as ordered mesoporous non-activated carbon, ordered mesoporous activated carbon and nitrogen-containing microporous carbon. During preparation, additional micropores formed in the interfacial regions between heterogeneous phases and greatly increased the specific surface area and porosity: when compared with the parent materials, the CO_2 take-up ability of the composites was indeed greatly improved due to the presence of micropores at the interfaces. The composite which comprised nitrogen-containing microporous carbon and Cu-BTC exhibited the highest CO_2 capacity: 8.24 and 4.51mmol/g under 1bar at 0 and 25C, respectively. In a natural extension of this idea, copper-containing metal-organic framework composites with graphene oxide have been synthesized[83] by using simple methods and tested with respect to the selective adsorption of CO_2 over N_2 under ambient conditions. The materials exhibited CO_2 up-takes of up to 9.59 and 5.33mmol/g at 0 and 25C, respectively, under a pressure of 1bar. A composite which comprised some 10wt% of graphene oxide exhibited the best CO_2 adsorption selectivity over N_2. Although most of the graphene-containing composites possessed a lower CO_2 adsorption ability, they exhibited a higher CO_2/N_2 selectivity than that of the base material. In similar work, Cu-BTC and graphene oxide composites were developed[84] by using a mixed-solvent method at 323K. The addition of N,N-dimethylformamide enabled the crystallization of Cu-BTC at low temperatures by accelerating nucleation. The product had a much higher surface area and total pore volume as compared with the usual Cu-BTC, yielding a CO_2 adsorption capacity of 8.02mmol/g at 273K under a pressure of 1bar; 17 to 90% higher than that of conventionally prepared Cu-BTC.

The photocatalytic conversion of carbon dioxide to methanol has been studied[85] by using Cu_3BTC_2. This metal-organic framework has a high surface area, and was used as a reactant in the preparation of a photocatalyst which contained TiO_2. The TiO_2-Cu_3BTC_2 photocatalyst was obtained by hydrolysing Ti^{IV} isopropoxide under solvothermal conditions. The TiO_2-Cu_3BTC_2 exhibited marked photocatalytic activity, as compared with TiO_2 nanoparticles, and the modified photocatalyst could be used to convert carbon dioxide into methanol by exploiting visible light. The metal-organic framework, $Cu_3(NH_2BTC)_2$, was prepared[86] using solvothermal methods, with amine-based trimesic acid as the organic ligand. The CO_2 adsorption was then studied in tests based upon a fixed-bed reactor, and the results showed that the amine groups had been successfully grafted onto $Cu_3(BTC)_2$. The CO_2 adsorption capacity increased to 1.41mmol/g at 10kPa and 50C, and this improvement in take-up was attributed to both physical and chemical adsorption. In other work, $Cu_3(BTC)_2$ was prepared[87] via electrochemical methods, and used for CO_2 and CH_4 adsorption; the greatest amounts of carbon dioxide and methane sorption being 26.89 and 6.63wt%, respectively, at 298K. The heat of adsorption of CO_2 decreased monotonically, while the reverse change occurred for CH_4. The selectivity of CO_2 over CH_4 also increased with increasing pressure and with the composition of the carbon dioxide component. Regeneration of the as-prepared material over 6 cycles led to no appreciable reduction in the CO_2 adsorption capacity. The $Cu_3(BTC)_2$ has also been introduced[88] into a carbon-paper based gas-diffusion electrode, showing that the faradaic efficiencies of CH_4 on gas-diffusion electrodes with Cu-metal-organic framework weight-ratios in the range of 7.5 to 10% were 2 to 3 times higher than those of gas-diffusion electrodes without Cu-framework additions under negative potentials of 2.3 to $2.5V_{SCE}$, while the faradaic efficiency of the competing hydrogen-evolution reaction was reduced to 30%.

The electrochemical carbon dioxide reduction reaction produces various chemical species, and copper clusters having suitably chosen surface coordination numbers provide active sites. One strategy[89] involves metal-organic framework-controlled Cu-cluster formation which shifts the CO_2 electroreduction towards multiple-carbon product generation. Under-coordinated sites are promoted during Cu-cluster formation by controlling the Cu-dimer structure. The symmetrical paddle-wheel Cu-dimer secondary building block of HKUST-1 is distorted into an asymmetrical form by separating adjacent benzene tricarboxylate components via heat treatment. The formation of Cu clusters, having a low coordination number, from distorted Cu dimers occurs in HKUST-1 during CO_2 electroreduction. Solvent-free preparation of HKUST-1 is an easy way to produce heterometallic metal-organic frameworks as electrocatalytic materials in the reduction of carbon dioxide. $Zinc^{II}$, $ruthenium^{III}$ and $palladium^{II}$ have been selected[90] as dopant metals to replace a portion of the $copper^{II}$ atoms of the original structure. Samples which were used as electrodes in a continuous-flow filter-press electrochemical cell resulted in activities as high as 47.2%, although such activity decreased with time. The performance of the Cu^{II}- and Bi^{III}-based metal-organic frameworks, HKUST-1 and CAU-17 respectively, with regard to the electroreduction of CO_2 to alcohols in such a filter-press cell has been investigated[91]. The materials were supported on porous carbon paper so as to form gas diffusion electrodes in 0.5M $KHCO_3$ aqueous solution and permit the continuous electrochemical conversion of CO_2 into methanol and ethanol, together with formic acid, hydrogen, carbon monoxide and ethylene. The maximum reaction rates and faradaic efficiencies for CO_2 conversion into methanol and ethanol were

$29.7 \mu mol/m^2 s/8.6\%$ and $48.8 \mu mol/m^2 s/28.3\%$, respectively, at a current density of $20 mA/cm^2$. This was an improvement of the values obtained for the copper- and bismuth-based materials separately. This synergistic effect was attributed to an interplay between the active sites and reaction intermediaries which promoted methanol formation, and a C-C coupling reaction to ethanol. The reaction-selectivity for alcohols could thus also be controlled by the Cu/Bi content of the electrode surface and by the current density. A 12% bismuth content was deduced to be the optimum level for the production of alcohols. In the case of the current density, CO_2 reduction was more favourable to methanol at $10 mA/cm^2$ while a level of $20 mA/cm^2$ made ethanol the predominant CO_2 reduction product. The performance of the combined frameworks remained essentially stable after 5h of operation.

Well-defined nanostructures in the sub-10nm range can be produced via the local chemical activation of surface-anchored metal-organic frameworks using an electron-beam lithographic technique. An investigation has been made[92] of the beam-induced surface activation of surface-anchored layers of HKUST-1 and copper[II] oxalate, with subsequent autocatalytic growth of deposits from $Fe(CO)_5$ and $Co(CO)_3NO$. Identification of the chemical species that triggered the decomposition of precursors on activated surfaces indicated that the electron-beam technique, as applied to HKUST-1, worked for $Fe(CO)_5$ but not for $Co(CO)_3NO$. Copper[II] oxalate did not respond to the electron-beam technique for either precursor, suggesting that copper nanoparticles are not active sites for the initiation of autocatalytic growth.

The HKUST-1 metal-organic framework can be modified by lithium doping in order to increase CO_2 adsorption. The optimum CO_2 adsorption capacity was obtained[93] by using a lithium nitrate solution of moderate concentration for doping, with the crystalline structure and morphology of HKUST-1 being retained together with its thermal stability. It was suggested that the incorporation of lithium increases the interaction between CO_2 molecules and the metal-organic framework. As a result, the CO_2 adsorption capacity was greatly increased.

A visible-light driven hybrid photocatalyst, consisting of gold nanoparticles, a Ti-substituted keggin-type polyoxometalate, $[PTi_2W_{10}O_{40}]^{7-}$, and HKUST-1 has been prepared[94] under atmospheric conditions with $[PTi_2W_{10}O_{40}]^{7-}$ acting as both an electron and a proton reservoir, with an active center in the HKUST-1 to boost CO_2 reduction, with HKUST-1 acting as a microreactor to concentrate CO_2 molecules and with gold nanoparticles harvesting visible light. As compared with $[PW_{12}O_{40}]^{3-}$ -containing HKUST-1 having an octahedral shape, the Ti-substituted equivalent had a cubic shape with a little corner cut. This was attributed to the higher net charge on the terminal oxygen atoms of $[PTi_2W_{10}O_{40}]^{7-}$, which slowed crystal growth in the <100> direction.

The $[PTi_2W_{10}O_{40}]^{7-}$ is always exposed on the {100} plane of HKUST-1 during synthesis, stabilizing gold nanoparticles and being evenly dispersed. Due to the stronger protonation of Ti-O and Ti-O-W, this material exhibited greater CO_2 reduction activity and selectivity under visible-light irradiation ($\lambda > 420nm$), by about 85.3 and 5.2 times, respectively; corresponding to CO_2-to-CO and CO_2-to-H_2 processes.

Grand canonical Monte Carlo simulations have been made[95] of the influence, of the framework's metal component upon CO_2/H_2-mixture adsorption and separation, in HKUST-1 and M-HATGUF, where M was Cd, Co, Cr, Cu, Fe, Mn, Mo, Ni, Ru or Zn. The results showed that the nature of the metal affected the CO_2 and H_2 take-ups. For example, Cr-HKUST-1 and Cd-HKUST-1 exhibited 11 and 38% enhanced CO_2/H_2 selectivities, together with 27 and 60% enhanced adsorbent performances respectively, as compared with Cu-HKUST-1. The CO_2/H_2 selectivity of HATGUF was almost doubled by changing the metal from Cu to Cd, together with a 142% increase in the adsorbent performance over that of HATGUF.

Mixed-matrix membranes which are formed by incorporating metal–organic frameworks into polymers suffer from a general limitation in that the former are usually 1-, 2- *or* 3-dimensional. The first offers the compensation of network percolation, the second offers access to high aspect-ratios and the third offers easy processing. It is preferable however to combine these factors. A simple method for forming multi-dimensional HKUST-1 nanoparticles has been proposed[96] which involves using a modulator to influence the metal-organic framework nucleation and growth. With 30wt% of multidimensional framework content, the mixed-matrix membrane exhibited a CO_2 permeability which was 2.5 times that of the pure polymer. There was also no great loss in selectivity for CO_2/CH_4 and CO_2/N_2, and almost no plasticization pressure response for CO_2 pressures of up to 750psi. Self-supported flexible HKUST-1 membranes with metal-organic framework contents of up to 82wt% can be produced[97] by a combination of electrospinning, multistep seeded growth and activation. The membrane exhibits a CO_2 adsorption capacity of 3.9mmol/g, good CO_2/N_2 selectivity and notable recyclability: the CO_2 capacity retains some 95% of its initial value after 100 adsorption-desorption cycles. An easy and quick process has been proposed[98] for shaping metal-organic frameworks at a concentration of less than 15%. Thus NbOFFIVE-1-Ni powders were shaped by using polymers as binders and their CO_2 capture properties were monitored before and after shaping. The polymer used was either rubbery (polyethyleneglycol) or glassy (polymethyl methacrylate). The method was also applied to the making of beads from UiO-66, ZIF-8 and HKUST-1. The results clearly showed that the glassy polymer was a better binder than was the rubbery one, given that it imparted high mechanical stability with relatively little change in gas adsorption at a 10% polymer content.

By combining a copper[II]-porphyrin zirconium metal-organic framework, PCN-224(Cu), and TiO_2 nanoparticles, a photocatalytic system was developed[99] which could markedly increase the catalytic activity of TiO_2 in the CO_2 photoreduction process. Without requiring a co-catalyst or a sacrifice reagent, the CO evolution-rate could attain 37.21µmol/gh. This was about 10 and 45.4 times higher than the rates for PCN-224(Cu) and pure TiO_2: 0.82 and 3.72µmol/gh, respectively. The improvement was attributed to increased light-harvesting due to the metalloporphyrin-based framework material, and to an interaction mechanism between the framework and the TiO_2 which favoured the separation of photo-excited charges. A method has been demonstrated[100] which can create crystalline copper-based metal–organic frameworks within 18min by using high-energy He–Ag lasers. The resultant *meso*-tetraphenyl-porphinato–Cu[II] frameworks had a Brunauer–Emmet–Teller specific surface area of 3418m^2/g, and exhibited a CO_2-capture adsorption capacity of 76.66mg$_{CO2}$/g within 27.5min at 25C and 40bar.

The cyclo-coupling of epoxides and CO_2 has been investigated[101] by using porphyrin-based Cu[II] metal-organic frameworks having 2-dimensional coordination networks. Various mono- and di-substituted epoxides could be transformed into cyclic carbonates under mild conditions. The catalytically active site was the di-copper paddle-wheel unit rather than the copper porphyrin complex component. The combination of a porphyrin-based copper metal-organic framework with tetrabutylammonium bromide efficiently catalyzed the cyclo-coupling of epoxides and CO_2.

Copper[II] species were loaded[102] into the large pores of an iron-based metal-organic framework with a large surface area and reduced them to Cu[I] species under mild conditions by utilizing Fe[II] sites in the pores. Remarkably, the Cu(I)-incorporated MOF (0.9Cu-MIL-100) exhibits a high CO/CO_2 selectivity (29 at 100kPa) and a large CO working capacity (1.61mmol/g at 10 to 100kPa) simultaneously, which has not been observed for previously reported adsorbent materials. Moreover, 0.9Cu-MIL-100 also presents very high CO/CH_4 and CO/N_2 selectivities (87 and 677). Furthermore, breakthrough and cyclic adsorption-desorption experiments confirm that this material can efficiently separate CO/CO_2 mixtures under dynamic mixture flow conditions and can be easily regenerated under mild conditions. This study provides a new strategy for developing adsorbents with both high CO/CO_2 selectivities and large CO working capacities.

A facile metal-organic framework-mediated strategy to obtain an efficient electrocatalyst for the synthesis of methane is suggested. Copper-based MOF-74 was chosen[103] as the precursor, which was electrochemically reduced to obtain copper nanoparticles. The porous structure of the MOF serves as a template for the synthesis of isolated copper nanoparticle clusters with high catalytic activities and high efficiencies for CH_4

production in the electrochemical CO_2 reduction reaction. The MOF-derived copper nanoparticles exhibited a high faradaic efficiency (>50%) for CH_4 production, with a 2.3-fold higher methanation activity at -1.3V_{RHE} as compared to that of commercial copper nanoparticles.

Monte Carlo simulations of gas sorption were performed[104] in Cu-TDPAH, also known as rht-MOF-9, a metal-organic framework with rht topology consisting of Cu^{2+} ions coordinated to 2,5,8-tris(3,5-dicarboxyphenylamino)-1,3,4,6,7,9,9b-hepta-azaphenalene ligands. This MOF is notable for the presence of open-metal copper sites and a high nitrogen content on the linkers. The material exhibited one of the highest experimental H_2 take-ups at 77K/1atm among the known rht-MOF family (about 2.72wt%) and also had a strong affinity for CO_2 (5.83mmol/g at 298K/1atm). The simulations, which included explicit many-body polarization interactions, accurately modelled macroscopic thermodynamic properties (e.g., sorption isotherms and isosteric heats of adsorption) as well as the binding sites for H_2, CO_2, CH_4, C_2H_2, C_2H_4 and C_2H_6 in the MOF. Four different binding sites were identified by analysis of the radial distribution function about the 2 chemically distinct Cu^{2+} ions, simulated annealing calculations and examination of the 3-dimensional histogram showing the sites of occupancy: (1) at the Cu^{2+} ion facing towards the center of the linker (CuL), (2) at the Cu^{2+} ion facing away from the center of linker (CuC), (3) nestled between three $[Cu_2(O_2CR)_4]$ units in the corner of the truncated tetrahedral (T-T_d) cage and (4) straddling the copper nuclei parallel to the axis of the Cu-Cu bond within the T-T_d cage. The low-loading (initial) binding site in the MOF was highly sensitive to the partial charges of the Cu^{2+} ions that were used for parametrization. It was noted that most sorbates prefer to sorb onto or near to the Cu^{2+} ions that exhibit the greater partial positive charge (i.e., at site 1). The simulated H_2 and CO_2 sorption results obtained using a polarizable potential for the respective sorbates were in good agreement with the corresponding experimental data; especially at near to ambient pressure.

The electroreduction of CO_2 to CO is much improved[105] by the design of copper-metal-organic-frameworks-derived nanoparticle (Cu-MOF/NP) catalysts, in which Cu/Cu_2O particles form a porous octahedral structure containing tunable Cu^0 and Cu^+ catalytic active sites. The ECR-CO_2 can be realized with a high current density of 25.15mA/cm^2 at a very low applied potential of only 0.79V_{RHE} even in an H-type cell, due to the high-surface-area porous structure with optimal surface chemistry of exposed Cu cations. Notably, a new flow electrochemical reactor integrated with a membrane electrode assembly is designed not only to reduce greatly the applied potential (~200mV) but also prompt the sensitivity of the reactor for identifying and quantifying reaction products. Accordingly, the Cu-MOF/NP catalyst enables an ultrahigh current density beyond 230mA/cm^2 at a low applied potential of -0.86V_{RHE} in the flow membrane electrode assembly reactor and the ethanol product (often undetectable in the traditional H-type cell) to be harvested.

A novel microporous metal-organic framework (FJU-44), with abundant accessible nitrogen sites on its internal surface, was constructed[106] from the tetrapodal tetrazole ligand tetrakis(4-tetrazolylphenyl)ethylene (H_4TTPE) and copper chloride. Notably, the CO_2 take-up capacity (83.4cm^3/g, at 273K and 1bar) in the activated FJU-44a is higher than most of tetrazolate-containing MOF materials. In particular, FJU-44a exhibited a superior adsorption selectivity of CO_2/N_2 (278 to 128) and CO_2/CH_4 (44 to 16), which was comparable to some well-known CO_2 capture materials.

Multifunctionalities, including hydrophobic methoxy groups, polar acylamide functionalities, and open copper^{II} sites, have been successfully integrated[107] into a twofold interpenetrated microporous MOF, HNUST-6. This possessed permanent porosity, with a moderate Brunauer–Emmett–Teller surface area of 1093m^2/g and a CO_2 adsorption capacity of 111cm^3/g at 1bar, with good selectivity for CO_2 over CH_4 (6.6) and N_2 (30.3) at 273K. This material exhibited an excellent water stability and its framework structure was retained following immersion in boiling water.

A highly porous (3,36)-connected txt-type acylamide-functionalized metal-organic framework (HNUST-8) was constructed from a pyridine-based acylamide-linking diisophthalate and dicopper^{II}-paddlewheel clusters[108]. Interestingly, HNUST-8 possesses an exceptionally water-stable framework with a Brunauer–Emmett–Teller surface area of about 2800m^2/g. At 298K, HNUST-8 exhibited a high excess CO_2 take-up of 19.7mmol/g at 40bar, an excellent total CH_4 storage capacity of 223cm^3(STP)/cm^3 with a large working capacity of 178cm^3(STP)/cm^3 at 80bar, as well as highly efficient CO_2/CH_4 and CO_2/N_2 separation under dynamic conditions at 1bar. Moreover, with the Lewis acidic open copper^{II} sites and Lewis basic acylamide groups integrated into the framework, HNUST-8 demonstrates efficient catalytic activity as an acid-base cooperative catalyst in a tandem one-pot de-acetalization-Knoevenagel condensation reaction.

Metal-organic framework-derived In-Cu bimetallic oxides are highly efficient electrocatalysts for the transformation of CO_2 into CO in an aqueous electrolyte. By controlling the In/Cu ratio, the faradaic efficiency of CO can reach 92.1%. The excellent performance[109] was attributed mainly to stronger CO_2 adsorption, a higher electrochemical surface area and a lower charge transfer resistance.

Theoretical calculations showed[110] that 3-coordinated copper sites have a higher activity than do 4-, 2- and 1-coordinated sites. A site-selective etching method was used to prepare a stacked-nanosheet metal–organic framework, CASFZU-1, based catalyst with a precisely controlled coordination number sites on its surface. The turnover frequency value of CASFZU-1 with 3-coordinated Cu sites, for the cyclo-addition reaction of CO_2 with epoxides, greatly exceeded those of previously reported catalysts. Five successive catalytic cycles reveal the superior stability of CASFZU-1 in a stacked-nanosheet structure.

A series of 2,5-alkoxybenzene-1,4-dicarboxylate ligands ($O_2CC_6H_2(OR)_2CO_2$, R = methyl-pentyl, 1 to 5, respectively) was used[111] to synthesize copper paddle-wheel MOFs. Rietveld and Pawley fitting of powder diffraction patterns for compounds Cu(3-5)(DMF) showed they adopted an isoreticular series with two-dimensional connectivity in which the interlayer distance increases from 8.68Å (R = propyl) to 10.03Å (R = pentyl). Adsorption of CO_2 by the MOFs was found to increase from 27.2 to 40.2cm^3/g with increasing chain length, which was attributed to the increasing accessible volume associated with increasing unit-cell volume. Ultrasound was used to exfoliate the layered MOFs to form MONs, with shorter alkyl chains resulting in higher concentrations of exfoliated material in suspension. The average height of MONs was investigated by AFM and found to decrease from 35 to 20nm with increasing chain length, with the thinnest MONs observed being only 5nm, corresponding to 5 framework layers.

The synthesis of two photoactive metal-organic frameworks, TCM-14 and TCM-15, was carried out[112] by incorporating 4,4′-azopyridine auxiliary ligands into pto-type scaffolds that were composed of dinuclear copper[II] so-called paddle-wheel based secondary building units and flexible, acetylene-extended, tritopic benzoate linkers. Room temperature CO_2 sorption of the MOFs was studied, and UV-light irradiation is shown to result in reduced CO_2 adsorption under static conditions. TCM-15 reveals a dynamic response leading to an instant desorption of up to 20% of CO_2 upon incidence of UV light because of the occurrence of non-periodic structural changes.

The assembly of mixed [1,1′3′,1″]terphenyl-4,5′,4″-tricarboxylic acid (H_3TPTC) and [1,1′-biphenyl]-4,4′-dicarboxylic acid (H_2BPDC), 2,2′-diamino-[1,1′-biphenyl]-4,4′-dicarboxylic acid (H_2BPDC-NH_2), or 6-oxo-6,7-dihydro-5H-dibenzo[d,f][1,3]diazepine-3,9-dicarboxylic acid (H_2BPDC-Urea) with Cu^{2+} ion generated the corresponding copper-paddlewheel-based metal-organic framework [Cu_5(TPTC)$_3$(BPDC)$_{0.5}$(H_2O)$_5$] (1), [Cu_5(TPTC)$_3$(BPDC-NH_2)$_{0.5}$(H_2O)$_5$] (1-NH_2) or [Cu_5(TPTC)$_3$(BPDC-urea)$_{0.5}$(H_2O)$_5$] (1-Urea). They are isostructural[113] with hierarchical porosity, consisting of zero-dimensional cage (19.2Å x 18.9Å) and 1-dimensional pillar channel (29.7Å x 15.1Å) in a manner of face-sharing. Analysis revealed that the porous volume ratios were 80.2, 80.0 and 77.8% for 1, 1-NH_2 and for 1-urea, respectively. Thermogravimetric measurements suggested 53, 51 and 48wt% guest molecules in 1, 1-NH_2, and 1-urea, respectively. 1-NH_2 and 1-urea were precisely functionalized via the introduction of amino and urea functional groups into the pillar channels. The constructed MOF 1-urea, incorporating both exposed copper active sites and accessible urea functional groups to substrates, presents high efficiency on catalytic CO_2 cyclo-addition with propene oxide to produce cyclic carbonate in the yield of 98% with a TOF value of 136/h at 1atm and room temperature.

A novel porous copper-based metal-organic framework $\{[Cu_2(TTDA)_2](DMA)_7\}_n$ (DMA = N,N-dimethylacetamide) was designed and synthesized[114] by combining a dual-functional organic linker 5'-(4-(4H-1,2,4-triazol-4-yl)phenyl)-[1,1':3',1''-terphenyl]-4,4''-dicarboxylic acid (H_2TTDA) and a dinuclear Cu^{II} paddle-wheel cluster. This metal-organic framework was characterized by elemental analysis, powder X-ray diffraction, thermogravimetric analysis, and single-crystal X-ray diffraction. The framework is constructed from two types of cage (octahedral and cuboctahedral) and exhibits two types of circular channel with an approximate size of 5.8 or 11.4Å along the crystallographic c-axis. Gas-sorption experiments indicated that it possessed a surface area of $1687m^2/g$ and a CO_2 adsorption capacity, at around room temperature, of up to $172cm^3/g$ at 273K and $124cm^3/g$ at 298K.

A modified slurry crystallization method was demonstrated[115] for making MOF copper 1,3,5-benzenetricarboxylate using a solvent-deficient system. One advantage was a marked reduction in solvent consumption and waste liquid. In a typical process, the mass ratio of ethanol to the solid reactants was about 0.52, which is only about 0.35 to 7.5% of that used in conventional processes. A yield of about 98.0% was easily achieved for a product with a uniform size up to 160μm. The obtained MOFs exhibited the characteristic microporous network with a surface area of about $1851m^2/g$ and a pore volume of ~$0.78cm^3/g$, which adsorbed about 6.73mol/kg of CO_2 at ordinary pressures. X-ray diffraction studies indicate that the MOFs possess an outstanding diffraction intensity ratio of the crystal plane (2,2,2) to (2,0,0), I(222)/I(200) = 22.4.

In order to explore the influence of modification sites of functional groups on the (CO_2/CH_4) gas separation performance of metal-organic frameworks, 6 types of organic linkers and 3 types of functional groups (i.e. -F, -NH_2, -CH_3) were used[116] to construct 36 metal-organic frameworks of pcu topology based upon copper paddle-wheels. Monte Carlo simulations were used to evaluate the separation performance of frameworks at low (vacuum swing adsorption) and high (pressure swing adsorption) pressures, respectively. Simulation results demonstrated that the CO_2 working capacity of the unfunctionalized frameworks generally exhibited a pore-size dependence at 1bar, which increased with decreasing pore size. It was also found that -NH_2 functionalized frameworks exhibited the highest CO_2 take-up, due to an enhanced Coulombic interaction between the polar -NH_2 groups and the quadrupole moment of CO_2 molecules, followed by -CH_3 and -F functionalized ones. Positioning of the functional groups, -NH_2 and -CH_3, at sites far from the metal node (site-b) led to a more significant improvement in CO_2/CH_4 separation performance, as compared with that adjacent to the metal node (site-a).

A ligand truncation strategy provides easy access to a wide variety of linkers for the construction of metal-organic frameworks having various structures and intriguing properties. This strategy was used[117] to design a novel bent di-isophthalate ligand, and to construct a copper-based framework, ZJNU-51, having the formula, $[Cu_2L(H_2O)_2] \cdot 5DMF$ (H_4L = 5,5'-(triphenylamine-4,4'-diyl) di-isophthalic acid). The ZJNU-51 was a 2-fold interpenetrated network in which the single network consisted of di-copper paddle-wheel units connected by the organic ligands and contained open channels as well as 6 distinct types of metal-organic cage. The gas-adsorption properties with respect to C_2H_2, CO_2 and CH_4 were systematically investigated, demonstrating that ZJNU-51 was a promising material for C_2H_2/CH_4 and CO_2/CH_4 separation. The adsorption selectivity at 298K and 1atm attained 35.6 and 5.4 for equimolar C_2H_2/CH_4 and CO_2/CH_4 gas mixtures, respectively.

A novel porous material, $[Cu(dpt)_2(SiF_6)]_n$, termed UTSA-120, where dpt is 3,6-di(4-pyridyl)-1,2,4,5-tetrazine, was developed[118] which was isoreticular to the net of SIFSIX-2-Cu-i. This material simultaneously exhibited a high CO_2 capture capacity (3.56mmol/g at 0.15bar and 296K) and CO_2/N_2 selectivity ($\sim$600), both of which were superior to those of SIFSIX-2-Cu-i and most other metal-organic framework materials. Neutron powder diffraction experiments revealed that the exceptional CO_2 capture capacity in the low-pressure region and the moderate heat of CO_2 adsorption could be attributed to the amenable pore size and dual functionalities (SiF_6^{2-} and tetrazine), which not only interacted with CO_2 molecules but also permitted the dense packing of CO_2 molecules within the framework. Simulated and real breakthrough experiments demonstrated that UTSA-120a could efficiently capture CO_2 gas from CO_2/N_2 (15/85, v/v) and CO_2/CH_4 (50/50) gas mixtures under ambient conditions.

Reaction of $CuSiF_6 \bullet xH_2O$ and 1,4-bis(4-pyridyl)-2-trifluoromethylbenzene (bpb-CF$_3$) through liquid diffusion produces[119] a porous SIFSIX-type metal-organic framework, $[Cu(bpb-CF_3)_2(SiF_6)]$ (UTSA-121) containing functional trifluoromethyl groups. Single-crystal X-ray diffraction analysis of UTSA-121 showed that bpb-CF$_3$ could well substitute for the prototypical dipyridine ligand to form a non-interpenetrating pcu framework, which was highly porous (void = 65.7%) and contained 3-dimensional intersecting channels with functionalized trifluoromethyl groups on the pore surface. The Brunauer-Emmett-Teller surface area of the activated UTSA-121a was up to 1081m^2/g. Gas sorption measurements of UTSA-121a revealed high C_2H_2 and CO_2 take-ups at room temperature. Ideal adsorbed solution theory calculations revealed that UTSA-121a exhibited highly selective adsorption of C_2H_2 and CO_2 over CH_4 and N_2 under ambient conditions.

A new approach to the fine-tuning of multinuclear clusters of metal-organic frameworks via symmetry-upgrading of isoreticular transformation was presented[120] and a bcu-type metal-organic framework, $\{[Cu_4(\mu_4\text{-}O)Cl_2(IN)_4][CuCl_2]\}_\infty$ (NJU-Bai35; NJU-Bai), with clusters $[Cu_4(\mu_4\text{-}O)Cl_2(COO)_4N_4]$ of higher symmetry compared to the pristine framework, was synthesized. Symmetry up-grading, implemented on the inorganic part, triggered the adjustment of channels in NJU-Bai35 to fit CO_2 molecules, leading to a high CO_2 adsorption capacity (7.20wt% at about 0.15bar and 298K) and a high selectivity of CO_2 over N_2 and CH_4 (275.8 for CO_2/N_2 and 11.6 for CO_2/CH_4) in NJU-Bai35.

Mixed-matrix membranes containing both 2-dimensional (ns-CuBDC) and 3-dimensional (ZIF-8) metal-organic frameworks were fabricated[121] in order to investigate their potential capabilities in CO_2/CH_4 separation. The mixed-matrix membrane containing ns-CuBDC alone was capable of improving CO_2/CH_4 selectivity whereas the mixed-matrix membrane containing ZIF-8 alone ensured improved CO_2 permeability. However, by combining both fillers in a mixed-matrix membrane aimed at tailoring the CO_2/CH_4 separation properties, both the CO_2 permeability and CO_2/CH_4 selectivity were improved by 16.6 and 30.5%, respectively, which indicated a highly desirable means of performance enhancement. Analysis of solubility-diffusivity in the mixed-matrix membranes revealed that the use of both fillers could improve both solubility selectivity and diffusivity selectivity. The overall results implied that the separation properties of gas separation membranes could be readily adjusted to meet the requirements of real-world applications by the combined use of two fillers of differing geometry.

A porous, Cu^{II}-metal organic framework (Cu-MOF) comprising a rigid lactam functionalized ditopic ligand (H_2L) was synthesized[122] at room temperature under slow evaporation conditions $\{H_2L = (5\text{-}(1\text{-}oxo\text{-}2,3\text{-}dihydro\text{-}1H\text{-}inden\text{-}2\text{-}yl)isophthalic acid)\}$. The single-crystal X-ray structure revealed the formation of a 3-dimensional framework of Cu-MOF with 1-dimensional channels decorated with lactam groups and exposed metal centers on the crystallographic c-axis. The Cu^{II}-coordinated DMF molecules were eliminated from the Cu^{II} metal center upon activation of Cu-MOF at a temperature of 150C under high vacuum to generate a solvent free framework with pores lined with unsaturated Lewis acidic Cu^{II} ions, i.e., Cu-MOF′. The lactam-functionalized channels inclined toward the CO_2, which interact with the Cu^{II} metal sites lining the channels of Cu-MOF′ and exhibit fascinating solvent-free heterogeneous catalytic conversion of CO_2 to cyclic carbonates at an atmospheric pressure of CO_2, under mild conditions.

Furthermore, the Cu-MOF' catalyst was easily recycled and reused for several cycles without any significant loss in catalytic activity.

Amine groups, being basic in nature, have an excellent affinity for the acidic CO_2 group. Amine-functionalized metal organic framework materials have promising characteristics as dry adsorbents for post-combustion CO_2 capture, owing to their enhanced CO_2 capture capacity. Work was focused on the synthesis[123] of a new amino-functionalized Cu-based MOF using 2-aminoterephthalic as a linker. Single-crystal X-ray diffraction studies revealed that an amine-functionalized metal-organic framework was created which had free -NH_2 groups aiding CO_2 adsorption. A CO_2 adsorption study indicated that the framework exhibited a better CO_2 capture tendency (5.85mmol/g at 25C) than did the Cu-BDC metal-organic framework.

Preparation methods were presented for thin dual-layer membranes and mixed-matrix membranes based upon nanosheets of Cu-BDC (lateral size-range 1 to 5µm, thickness of 15nm) and commercially available poly(ethylene oxide)–poly(butylene terephthalate) (PEO–PBT) copolymer (Polyactive™) and their performances were compared with regard to CO_2/N_2 separation. The mixed-matrix membranes and dual-layer membranes represented two extremes, on the one hand with well-mixed components and on the other hand completely segregated layers. Compared to the free-standing membranes, the thin PAN- and zirconia-alumina-supported mixed-matrix membranes showed significant enhancement in both permeance and selectivity. The support properties affected the obtained selective layer thickness and its resistance affected[124] the CO_2/N_2 selectivity. The permeance of thin dual-layer membranes was among the highest reported for framework-based thin mixed-matrix membranes, but had a modest selectivity. Addition of the nanosheets to the thin membranes further improved the CO_2/N_2 selectivity, of the already selective polymer, to 77. The nanosheets in the thin membranes rendered a gutter-layer on the PAN support superfluous. The small pore support ZrO_2-alumina did not need a gutter-layer. X-ray diffraction revealed that the spatial distribution of metal-organic framework nanosheets and polymer chains packing were responsible for differences in the permeation performance of the free-standing, thin dual layer and mixed-matrix membranes.

A microdroplet-based spray process has been used[125] for the fast and easy synthesis of bimetallic metal–organic frameworks, including Cu-TMA(Fe), Cu-TMA(Co), Cu-TMA(Mg), Cu-TMA(Al) and Cu-TPA(Fe), where TMA is trimesic acid and TPA is terephthalic acid. By forming M-O (M: metal, O: oxygen) bonds with the ligands, the secondary metal sites were incorporated into the framework, partially replacing the original Cu sites in the parent MOF (i.e., Cu-TMA), which induced changes in surface areas, pore structures and gas sorption properties. In particular, the bimetallic MOFs

Materials Research Forum LLC
https://doi.org/10.21741/9781644900857

exhibited slightly higher surface areas than did the parent MOF; which was attributable to the expansion of the unit cells, specifically because of the elongated M-O bonds. Meanwhile, the pore volumes of the MOFs showed a positive correlation with the atomic radii of the metal sites, suggesting that the metal substitution is an effective method to adjust the pore structures of the MOFs. In addition, various gas sorption performances were observed among the bimetallic MOFs and the parent MOF, which was mainly associated with the differing electrostatic interaction between the gas molecules and the frameworks induced by the incorporation of various secondary metal sites. Specifically, metal sites with a larger electronegativity have a higher impact on the properties of the adsorbed CO_2, such as C=O bond length and O=C=O bond angle, leading to a more asymmetrical geometry and polarization of the adsorbed CO_2 molecules. As a result, the gas sorption capacity, selectivity and isosteric heat of adsorption varied with various secondary metal sites within the framework.

A novel 3-dimensional metal-organic framework, BSF-1, based upon the closo-dodecaborate cluster $[B_{12}H_{12}]^{2-}$ was readily prepared[126] at room temperature by supramolecular assembly of $CuB_{12}H_{12}$ and 1,2-bis(4-pyridyl)acetylene. The permanent microporous structure was studied by X-ray crystallography, powder X-ray diffraction, IR spectroscopy, thermogravimetric analysis, and gas sorption. The experimental and theoretical study of the gas sorption behavior of BSF-1 for N_2, C_2H_2, C_2H_4, CO_2, C_3H_8, C_2H_6 and CH_4 indicated excellent separation selectivities for C_3H_8/CH_4, C_2H_6/CH_4, and C_2H_2/CH_4 as well as moderately high separation selectivities for C_2H_2/C_2H_4, C_2H_2/CO_2, and CO_2/CH_4. Moreover, the practical separation performance of C_3H_8/CH_4 and C_2H_6/CH_4 was confirmed by dynamic breakthrough experiments.

A new series of flexible ligands derived from p-xylylenediamine, which vary in the positioning of carboxylic acid functional groups, have been found[127] to form a series of coordination polymers and a discrete coordination complex. Use of the tetra-substituted meta-carboxylic acid ligand N,N,N′,N′-tetra(3-carboxybenzyl)-p-xylylenediamine, isolated as its dihydrochloride salt (H_6M_4pxy)Cl_2, resulted in the simultaneous formation of two products using the same synthetic conditions. One of these products was a discrete cage-type coordination complex of the form $[Cu_4(M_4pxy)_2(DMF)_2(OH_2)]$ (1), while the other product was the 2-dimensional coordination polymer poly-$[Cu_2(M_4pxy)(OH_2)_2]$ (2). Synthesis utilising the tetra-substituted para-carboxylic acid ligand, isolated as N,N,N′,N′-tetra(4-carboxybenzyl)-p-xylylenediamine dihydrochloride (H_6P_4pxy)Cl_2, resulted in the formation of multiple 2-dimensional coordination polymers, poly-$[Cu_2(P_4pxy)(DMF)(OH_2)]\cdot2DMF\cdot3H_2O$ ($3\cdot2DMF\cdot3H_2O$), poly-$[Mn_2(P_4pxy)(DMA)_2]$ (4), poly-$[Co_4(P_4pxy)_2(DMF)_2(\mu-OH_2)_2(OH_2)_2]\cdot6DMF\cdot4H_2O$ ($5\cdot6DMF\cdot4H_2O$) and poly-$[Cd_2(P_4pxy)(OH_2)_2]\cdot DMA\cdot3H_2O$ ($6\cdot DMA\cdot3H_2O$). The last of these materials, a 3-

dimensional coordination polymer, was found to adsorb some $32cm^3/g$ STP of CO_2 at atmospheric pressure and 273K.

By combining first-principles calculations and molecular simulations, an atomistic-level explanation was provided[128] for a marked change in CO_2 adsorption upon light treatment in two photo-active metal–organic frameworks: PCN-123 and $Cu_2(AzoBPDC)_2(AzoBiPyB)$. It was demonstrated that the reversible decrease in gas adsorption upon isomerization can be attributed mainly to blocking of the strongly adsorbing sites at the metal nodes by azobenzene molecules in a cis configuration. The same mechanism was also found to apply to molecules such as alkanes and toxic gases. For example, the metal node-blocking mechanism can be leveraged to achieve optimal adsorption properties as a function of metal substitution and/or ligand functionalization. The working capacity could be increased by a factor of two in PCN-123 by replacing the Zn_4O node with the more strongly adsorbing Mg_4O.

It is challenging to synthesize 2-dimensional MOFs with controllable size and functionalities using direct and scalable approaches under mild conditions (e.g., room temperature). One-step room-temperature synthesis was used to prepare a series of 2-dimensional MOFs based upon Cu^{II} paddle-wheel units, where the intrinsically anisotropic building blocks led to the anisotropic growth[129] of 2-dimensional MOF nanoparticles, and the pillared structure led to high surface areas. The size of 2-dimensional MOFs could be adjusted by using a DMF/H_2O mixed solvent. The thinnest particles were around 3nm thick, and the highest aspect ratio was up to 200. The functionalization of 2-dimensional MOFs was also achieved by selecting ligands with desired functional groups. The gas sorption results revealed that amino and nitro-functionalized 2-dimensional MOFs exhibited a higher CO_2 sorption selectivity over CH_4 and N_2, suggesting that these materials can be further applied to natural-gas sweetening (CO_2/CH_4 separation) and carbon capture from flue gas (CO_2/N_2 separation).

A 3-dimensional microporous metal-organic framework with open Cu^{2+} sites and suitable pore space, $[Cu_2(L)(H_2O)_2]\cdot(H_2O)_4(DMF)_8$ (ZJU-15, H_4L = 5,5′-(9H-carbazole-2,7-diyl)diisophthalic acid; DMF = N,N-dimethylformamide; was constructed[130]. The activated ZJU-15a had 3 different types of cage, a Brunauer–Emmett–Teller surface area of $1660m^2/g$ and could separate gas mixtures of C_2H_2/CH_4 and CO_2/CH_4 at room temperature.

A new ligand, 1,3-bis(4-carboxyphenyl)-4,5-dihydro-1H-imidazol-3-ium tetrafluoroborate, $H_2Sp5-BF_4$, has been used[131] for the construction of a novel MOF, Cu-Sp5-EtOH. This highly crystalline material has a charged framework that is expected to give rise to a high CO_2/N_2 selectivity. However, the pores of the parent structure could not be accessed due to the presence of strongly coordinated ethanol molecules. After solvent exchange with methanol and subsequently heating Cu-Sp5-MeOH under vacuum, it was possible to liberate the solvent, providing other small molecules such as CO_2 with access to the inside of the now-porous structure, Cu-Sp5. The combination of open metal sites and framework charge led to an exceptionally high CO_2/N_2 selectivity, as determined by ideal adsorbed solution theory calculations performed on single-component adsorption isotherms. The CO_2/N_2 selectivity of Cu-Sp5 reaches a value of over 200 at pressures typically found in post-combustion flue gas (0.15bar CO_2/0.85bar N_2), a value that is among the highest reported to date.

Europium-

Three new lanthanide-based metal-organic frameworks, MOF-590, -591 and -592 constructed from a tetratopic linker, benzoimidephenanthroline tetracarboxylic acid (H4BIPA-TC), were synthesized[132] under solvothermal conditions. All of the new metal-organic frameworks had 3-dimensional frameworks, which had previously-unobserved topologies. Gas-adsorption measurements of MOF-591 and MOF-592 revealed a good adsorption of CO_2 (low pressure, room temperature) and moderate CO_2 selectivity over N_2 and CH_4. Breakthrough experiments subsequently illustrated the separation of CO_2 from binary mixtures of CO_2 and N_2 with the use of MOF-592. Accordingly, MOF-592 revealed selective CO_2 capture effectively without any loss in performance after 3 cycles. Moreover, MOF-590, MOF-591 and MOF-592 were catalytically active in the oxidative carboxylation of styrene and CO_2 for the one-pot synthesis of styrene carbonate under mild conditions (1atm CO_2, 80C, no solvent). Among the new materials, MOF-590 revealed a remarkable efficiency, with exceptional conversion (96%), selectivity (95%) and yield (91%).

Photocatalytic reduction of CO_2 is a promising approach to achieving solar-to-chemical energy conversion. However, traditional catalysts usually suffer from low efficiency, poor stability, and selectivity. It was demonstrated[133] that a large porous and stable metal-organic framework featuring dinuclear Eu^{III}_2 clusters as connecting nodes and Ru(phen)$_3$-derived ligands as linkers could be constructed to catalyze visible-light-driven CO_2 reduction. Photo-excitation of the metalloligands initiates electron injection into the nodes, to generate dinuclear Eu^{II}_2 active sites, which can selectively reduce CO_2 to formate in a two-electron process at the remarkable rate of 321.9μmol/h $mmol_{MOF}$. The electron transfer from Ru metalloligands to Eu^{II}_2 catalytic centers were studied via transient absorption and theoretical calculations and shed light on the photocatalytic mechanism.

Hafnium-

New acetylenedicarboxylate (ADC) and chlorofumarate (Fum-Cl) based hafnium-containing metal-organic frameworks have been synthesized[134] by alternately reacting acetylenedicarboxylic acid in N,N-dimethylformamide or water, with acetic acid as modulator. These materials, $[Hf_6O_4(OH)_4(ADC)_6]$ and $[Hf_6O_4(OH)_4(Fum-Cl)_6]$, were isostructural with UiO-66 and consisted of octahedral $[Hf_6O_4(OH)_4]^{12+}$ units, connected to others by 12 acetylenedicarboxylate or chlorofumarate linkers to give a microporous network. The $[Hf_6O_4(OH)_4(Fum-Cl)_6]$ adsorbed CO_2, with an isosteric heat of 39kJ/mol.

Copper nanoparticles, combined with zirconium/hafnium-based metal-organic framework materials, are effective in the hydrogenation of CO_2 to methanol. Among $Cu/\gamma-Al_2O_3$, Cu/ZIF-8, Cu/MIL-100 and Cu/UiO-66 composites, UiO-66 is the most active support, with Cu/Zr-UiO-66 producing methanol at a rate which is 70 times higher than that of $Cu/\gamma-Al_2O_3$. The replacement of Zr^{4+} by Hf^{4+} in UiO-66 tripled the rate of methanol production[135]. There is a further possible source of improvement in that Cu/Zr-UiO66-COOH yields a 3-fold increase in methanol production, as compared with that of Zr-UiO-66 or Zr-UiO66-NH$_2$. The increased catalytic activity of copper nanoparticles is attributed to the degree of charge transfer from copper nanoparticles to the UiO-66.

Table 6. Cyclo-addition of CO_2 and epichlorohydrin using various catalysts

Catalyst	Yield (%)
tetrabutylammonium bromide	16.2
$[CuL(ClO_4)_2]$ and tetrabutylammonium bromide	90.2
Hf-VPI-100(Cu) and tetrabutylammonium bromide	96.4
Hf-VPI-100(Ni) and tetrabutylammonium bromide	88.4

Two isostructural hafnium-based metal-organic frameworks, Hf-VPI-100(Cu) and Hf-VPI-100(Ni) have been investigated[136] as highly efficient catalysts for the cyclo-addition of CO_2 to epoxides at ambient pressures. The 3-dimensional porous metal-cyclam-based zirconium metal-organic frameworks were prepared[137] by assembling 8-connected Zr_6 clusters and metallo-cyclams as organic linkers. The cyclam core had accessible axial coordination sites for guest interactions, and retained the electronic properties of the parent cyclam ring. The VPI-100 frameworks exhibited excellent chemical stability in organic and aqueous solvents over wide pH ranges and a CO_2 take-up of up to about

9.83wt% adsorption at 273K under 1atm. The crystal structures of the frameworks were retained under CO_2 pressures of up to 20bar. A comparison of the catalytic efficiency of Hf-VPI-100 and of zirconium-based VPI-100 revealed that open metal centers in the metallocyclam acted as primary Lewis acid sites for aiding the catalytic conversion of CO_2. Tetrabutylammonium bromide was used as a co-catalyst at relatively low (1.5bar) CO_2 pressures. In a typical catalysis test, Hf-VPI-100 powder (0.008mmol), tetrabutylammonium bromide (0.31mmol) and epoxide (31.3mmol) were reacted in a 20ml stainless-steel autoclave at 90C under 1.5bar of CO_2 for 6h. Epichlorohydrin and 1,2-epoxybutane were used as epoxide substrates because of their low volatility. The Hf-VPI-100 exhibited high catalytic activity in the cyclo-addition of epichlorohydrin and CO_2. Under quite mild conditions (1.5bar CO_2, 90C, 6h), the yields were 95 and 90% for the copper and nickel analogues, respectively. Control experiments involving no metal-organic framework catalyst, and with only tetrabutylammonium bromide present, yielded just 16.2%. In the absence of tetrabutylammonium bromide, no conversion via the cyclo-addition reaction was detected. The Hf-VPI-100(Cu) out-performed the Cu^{II}-cyclam ligand, which gave a 90% yield of cyclic carbonate under the same reaction conditions.

Table 7. Key diffusivities in epoxide/VPI-100 reactions

Metal-Organic Framework	D_1 (cm²/s)	D_2 (cm²/s)
Hf-VPI-100(Cu)	5.2×10^{-10}	1.1×10^{-12}
Hf-VPI-100(Ni)	7.4×10^{-10}	1.8×10^{-12}
Zr-VPI-100(Cu)	4.7×10^{-11}	4.9×10^{-13}
Zr-VPI-100(Ni)	6.5×10^{-10}	1.9×10^{-12}

The frameworks could be easily separated from the reaction and reused, with no great loss of catalytic activity. Because the yields for epichlorohydrin were close to 100%, an investigation was made of CO_2 chemical fixation with 1,2-epoxybutane under the same conditions (1.5bar CO_2, 90C, 6h). Using this substrate, zirconium-based VPI-100 led to less catalytic conversion as compared with epichlorohydrin. A similar decrease in yield, as compared with that for epichlorohydrin, was observed when using the Hf-VPI-100 analogues. The Hf-VPI-100(Cu) and Hf-VPI-100(Ni) exhibited a higher catalytic activity than that of tetrabutylammonium bromide with regard to the formation of butylene carbonate; yielding 43.8 and 39.4%, respectively. Comparable conversion yields were obtained when using Zr-VPI-100 under the same reaction conditions, even though the hafnium node was more oxophilic than was the zirconium node. There were 2 different diffusion regimes. A faster sorption mode was characterized by a diffusion coefficient, D_1, which accounted for nearly two-thirds of the total take-up in only 90s. The slower sorption mode was characterized by a diffusion coefficient, D_2, which covered most of the remaining take-up during a period of about 1h (table 6). The diffusion in the faster sorption regime (D_1) was over 100 times faster (table 7) than that in the slower sorption regime (D_2). The Zr-VPI-100(Cu) system exhibited much slower diffusion than did other systems, in both sorption regimes (figure 4).

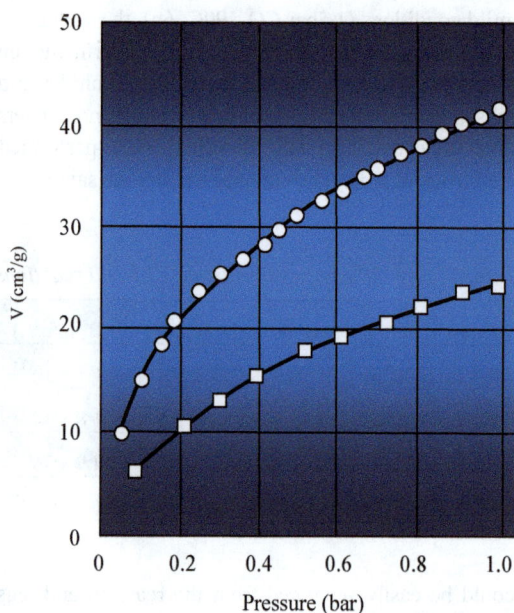

Figure 4. CO_2 adsorption isotherms for Hf-VPI-100(Cu)
circles: 273K, squares: 296K

Materials Research Forum LLC
https://doi.org/10.21741/9781644900857

Indium-

Four new 2-dimensional indium metal-organic frameworks, $(Me_2NH_2)[In(SBA)_2]$ (1), $(Me_2NH_2)[In(SBA)(BDC)]$ (2), $(Me_2NH_2)[In(SBA)(BDC-NH_2)]$ (3), and $(NH_4)_3[In_3Cl_2(BPDC)_5]$ (4), $(H_2SBA$ = 4,4′-sulfonyldibenzoic acid; H_2BDC = 1,4-benzenedicarboxylic acid; $H_2BDC-NH_2$ = 2-amino-1,4-benzenedicarboxylic acid; H_2BPDC = 4,4′-biphenyldicarboxylic acid) have been synthesized[138] under solvothermal reaction conditions for compounds 1 to 3 and the DES (deep eutectic solvent) reaction has been attempted for compound 4. The structure of these MOFs has been determined by using single-crystal X-ray diffraction studies for all of these four 2-dimensional monolayer framework with porous properties. The N_2 gas sorption measurements indicated that the Brunauer-Emmer-Teller and Langmuir surface areas of compound 1 are 207 and 301m^2/g, respectively, which is probably the first one, to date, to exhibit substantial gas take-up properties among the entire 2-dimensional In-MOF family. Furthermore, these new indium MOFs upon adding n-Bu4NBr were active for the cyclo-addition of CO_2 and propylene oxide; generating propylene carbonates in high conversions volumes under mild conditions. In particular, the most active MOF 4 was found to couple CO_2 efficiently with a series of terminal epoxides to give the corresponding cyclic organic carbonates with high selectivity.

The targeted synthesis of metal-organic frameworks with open metal sites, following reticular chemistry rules, provides a straightforward means for the development of advanced porous materials; especially for gas storage/separation. Using a palladated tetracarboxylate metalloligand as a 4-connected node, the first heterobimetallic In^{III}/Pd^{II}-based MOF with square-octahedron (soc) topology was prepared[139]. The new MOF, formulated as $[In_3O(L)1.5(H_2O)_2Cl]_n$(solv) (1), features the oxo-centered trinuclear clusters, $[In_3(\mu_3-O)(-COO)_6]$, acting as trigonal-prismatic 6-connected nodes that linked together with the metalloligand trans-$[PdCl_2(PDC)_2]$ (L4-) (PDC: pyridine-3,5-dicarboxylate) to form a 3-dimensional network. After successful activation of 1 using supercritical CO_2, high-resolution microporous analysis revealed the presence of small micropores (5.8Å) with a Brunauer–Emmett–Teller area of 795m^2/g and total pore volume of 0.35cm^3/g. The activated solid shows high gravimetric (92.3cm^3/g) and volumetric (120.9cm^3/cm^3) CO_2 take-up at 273K and 1bar as well as high CO_2/CH_4 (15.4 for a 50:50 molar mixture) and CO_2/N_2 (131.7 for a 10:90 molar mixture) selectivity, with a moderate Qst for CO_2 (29.8kJ/mol). Slight modifications of the synthesis conditions led to the formation of a different MOF with an anionic framework, having the chemical formula, $[Me_2NH_2][In(L)]_n$(solv). This MOF is constructed from pseudotetrahedral, mononuclear $[In(-COO)_4]$ nodes bridged by four L4- linkers, resulting in a 3-dimensional network with pts topology.

Solvothermal reactions between indium nitrate and two mixed-linkers produced[140] two new isoreticular 8-connected trinuclear indium-based AFMOFs of $[(In_3O)(OH)(L2)_2(IN)_2]\cdot(solv)x$ (2-In) and $[(In_3O)(OH)(L_2)_2(AIN)_2]\cdot(solv)x$ (NH_2-2-In) (H_2L2 = 4,4'-(carbonylimino) dibenzoic acid and HIN = isonicotinic acid or HAIN = 3-aminoisonicotinic acid), respectively. Moreover, by means of reticular chemistry, an extended network of $[(In_3O)(OH)(L3)_2(PB)_2]\cdot(solv)x$ (3-In) (H_2L3 = 4,4'-(terephthaloylbis(azanediyl))dibenzoic acid, HPB = 4-(4-pyridyl)benzoic acid) was also successfully produced after prolongation of the former dicarboxylate linker and HIN, resulting in a truly 8-connected isoreticular AFMOF platform. These frameworks were structurally determined by single-crystal X-ray diffraction. Sorption studies further demonstrate that 2-In and NH_2-2-In exhibit not only high surface areas and pore volumes but also relatively high carbon-capture capabilities (the CO_2 take-ups reach 60.0 and 75.5cm^3/g at 298K and 760 torr, respectively) due to the presences of amide and/or amine functional groups. The calculated selectivities of CO_2/N_2 and CO_2/CH_4 were 10.18 and 12.43, 4.20 and 4.23 for 2-In and NH_2-2-In, respectively, which were additionally evaluated by mixed-gas dynamic breakthrough experiments. High-pressure gas-sorption measurements showed that both materials could take up moderate amounts of natural gas.

Zeolite-like metal-organic frameworks are a distinct sub-set which possess zeolitic topologies but also exhibit an absence of self-interpenetration of 3-dimensional networks and make available larger pore spaces. They also exhibit excellent chemical stability in aqueous media and interesting ion-exchange possibilities. Such a zeolite-like framework material, $[(CH_3)_2NH_2][In(ABTC)]\bullet 3DMF$, has been synthesized[141] by using H_4ABTC (2,2',5,5'-azobenzenetetracarboxylic acid) as a linker to coordinate 4 $[In(O_2C)_4]^-$ secondary building units and generate a 3-dimensional framework. The present material offered great advantages in terms of chemical stability, and a N_2 adsorption study of activated samples indicated Langmuir and Brunauer–Emmett–Teller surface areas of 1302 and 966m^2/g, respectively. Among the metal ions, In^{III} has a high coordination number that tends to lead to the formation of stable indium-organic frameworks. The ion forms 3 types of secondary building unit in the latter frameworks: linear neutral $[In(\mu-OH)(O_2C^-)_2]$, charged trimeric $[In_3(\mu_3-O)(O_2C^-)_6]$ and tetrahedral $[In(O_2C^-)_4]/[InN_4(O_2C^-)_4]$. The In^{III} ion was thus selected here as the metal center and the so-called 4+4 strategy was used for assembly of the In-ZMOF via 4-connected tetrahedral $[In-(O_2C)_4]^-$ and tetrahedral organic ligands. The tetrahedral organic ligand node was attributed to the distorted 2,2',5,5'-azobenzenetetracarboxylic acid ligand, in which 2 benzene rings are almost perpendicular to one another. This novel In-ZMOF material exhibited good chemical stability and a CO_2 capture capability of 129cm^3/g at 273K under 1atm pressure (figure 5).

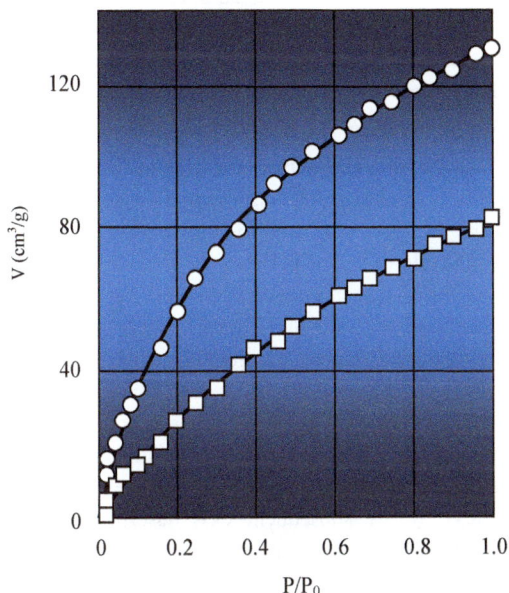

Figure 5. CO_2 adsorption isotherms for [(CH$_3$)$_2$NH$_2$][In(ABTC)]•3DMF under 1atm circles: 273K, squares: 298K

Iron-

An iron^{III}-based metal-organic framework was prepared[142] using solvothermal methods with tritopic bridging 1,3,5-tris(4-carboxyphenyl) benzene as a linker between the metal centers. Thermogravimetric analysis revealed that weight losses over the entire temperature range corresponded to the loss of adsorbed gases from pores and cavities, removal of N, N-dimethyl formamide and framework disintegration. The specific surface area was 300m^2/g, with very little increase occurring with partial pressure; indicating a microporous character. The maximum CO_2 adsorption capacity at room temperature and at a partial pressure of 0.9atm was 3.0wt%. A temperature increase from 298 to 343K decreased the adsorption capacity. Linearity of the CO_2 adsorption was attributed to weak adsorption sites.

A time-resolved diffraction analysis was made[143] of the kinetics of formation of a robust MOF, MFM-300(Fe), which has a high adsorption capacity for CO_2 (9.55mmol/g at 293K and 20bar). Applying the Avrami-Erofeev and the 2-step kinetic Finke-Watzky models to *in situ* high-energy synchrotron X-ray powder diffraction data obtained during the synthesis of MFM-300(Fe) permitted determination of the overall activation energy of formation (50.9kJ/mol), the average energy of nucleation (56.7kJ/mol), and the average energy of autocatalytic growth (50.7kJ/mol). The synthesis of MFM-300(Fe) has been scaled-up 1000-fold, enabling the successful breakthrough separations of the CO_2/N_2 mixture in a packed-bed with a selectivity for CO_2/N_2 of 21.6.

The gate-opening adsorption mechanism and sigmoidal adsorption isotherm were theoretically investigated[144] taking CO_2 adsorption into porous coordination polymers, $[Fe(ppt)_2]_n$ (PCP-N, Hppt = 3-(2-pyrazinyl)-5-(4-pyridyl)-1,2,4-triazole) and $[Fe(dpt)2]_n$ (PCP-C, Hdpt = 3-(2-pyridinyl)-5-(4-pyridyl)-1,2,4-triazole) as examples, where the hybrid method consisting of dispersion-corrected DFT for infinite porous coordination polymer and a post-Hartree-Fock (SCS-MP2 and CCSD(T)) method for the cluster model was employed. PCP-N has site I (one-dimensional channel), site II (small aperture to site I), and site III (small pore) available for CO_2 adsorption. The CO_2 adsorption at site I occurs in a one-by-one manner with a Langmuir adsorption isotherm. CO_2 adsorption at sites II and III occurs through a gate-opening adsorption mechanism, because the crystal deformation energy (EDEF) at these sites is induced largely by the first CO_2 adsorption but induced much less by the subsequent CO_2 adsorption. Interestingly, nine CO_2 molecules are adsorbed simultaneously at these sites because a large EDEF cannot be overcome by adsorption of one CO_2 molecule but can be by simultaneous adsorption of nine CO_2 molecules. For such CO_2 adsorption, the Langmuir-Freundlich sigmoidal adsorption isotherm was derived from the equilibrium equation for CO_2 adsorption. A very complicated CO_2 adsorption isotherm, experimentally observed, is reproduced by combination of the Langmuir and Langmuir-Freundlich adsorption isotherms. In PCP-C, CO_2 adsorption occurs only at site I with the Langmuir adsorption isotherm. Sites II and III of PCP-C cannot be used for CO_2 adsorption because a very large EDEF cannot be overcome by the simultaneous adsorption of nine CO_2 molecules. Factors necessary for the gate-opening adsorption mechanism were explained on the basis of differences between PCP-N and PCP-C.

The CO_2 conversion performance of 2-dimensional sheet metal-organic frameworks, such as $TM_3(HAB)_2$ (TM = Fe, Co, Ni, Cu; HAB = hexaaminobenzene), was studied[145] by combining density functional theory and a hydrogen electrode model. The $Fe_3(HAB)_2$ material was promising for use in the electrocatalytic CO_2 reduction reaction. Reaction-energy calculations identified the preferred pathway for CO_2 conversion to CH_3OH on $Fe_3(HAB)_2$ and indicated a corresponding free-energy change of 0.69eV, with an activation energy of 1.36eV. The formation of CHO via the hydrogenation of CO was the rate-limiting step. The catalytic activity of $Fe_3(HAB)_2$ could even be better than that of copper; the usual benchmark.

The metal-organic framework, Fe-BTC, was incorporated into a polymeric membrane with a view to achieving CO_2/N_2 gas separation[146]. The Fe-BTC was first characterized by making single-component adsorption equilibrium measurements of CO_2 and N_2 in order to evaluate its performance as an adsorbent for CO_2/N_2 separation. The Fe-BTC was then incorporated into Matrimid®5218 polymer at various percentages and pure gas permeation experiments were performed using CO_2 and N_2, at 303, 323 and 353K, in order to evaluate the effect of temperature upon permeability and CO_2/N_2 selectivity. Increased CO_2 permeability and CO_2/N_2 selectivity were in fact observed, especially at 353K.

A pure-supramolecular-linker approach to the formation of metal-organic frameworks was initially given, which was demonstrated[147] by the synthesis of two highly connected and isostructural metal-organic frameworks, $\{Fe_3O(TPBTM6-)(Cl)(H_2O)_2\}_\infty$ (TPBTM = N,N',N''-tris(isophthalyl)-1,3,5-benzenetricarboxamide) (NJU-Bai52,) and $\{Sc_3O(TPBTM6-)(OH)(H_2O)_2\}_\infty$ (NJU-Bai53). They exhibited exceptional thermal stability, water stability and highly selective CO_2 capture properties. In particular, NJU-Bai53 with higher take-ups (2.74wt% at 0.4mbar and 298K, 7.67wt% at 298K and 0.15bar) and higher selectivity may be an excellent candidate for CO_2 capture.

An electrocatalyst for CO_2 reduction in aqueous solution has been created[148] by combining 5,10,15,20-tetrakis(4-carboxyphenyl)porphyrinato-Fe^{III} chloride with UiO-66. The latter's unique framework provides protons which improve the CO_2 reduction activity on the iron porphyrin, leading to a faradaic efficiency of nearly 100% at an overpotential of 450mV when turning CO_2 into CO. Improving the stability of lead halide perovskite quantum dots in the presence of water is an important factor in ensuring their practical application[149]. To this end, $CH_3NH_3PbI_3$ (MAPbI3) perovskite quantum-dots have been encapsulated within the pores of the Fe-porphyrin-based metal organic framework, PCN-221(Fe), by sequential deposition so as to construct a series of composite photocatalysts[150]. The latter exhibited a much improved stability in systems containing water. Close contact of the quantum-dots with Fe catalytic sites in the metal-organic framework permits photogenerated electrons in the dots to transfer rapidly to the Fe catalytic sites and increase the photocatalytic activity for CO_2 reduction. When using water as an electron source, one of the composites exhibited a total yield of 1559µmol/g for photocatalytic CO_2 reduction to CO (34%) and CH_4 (66%). This is 38 times higher than that for the framework material without perovskite quantum-dots. Mesoporous nitrogen-doped carbon nanoparticles with atomically-dispersed iron sites have been created[151] by pyrolysis of the Fe-containing ZIF-8 framework material. The hydrolysis of tetramethyl orthosilicate in the framework before pyrolysis is essential to ensuring a high final surface area and impeding the formation of iron oxide nanoparticles. Experimental data, combined with theoretical calculations, suggest that the iron possesses an associated coordination sphere comprising a porphyrinic environment and OH/H_2O components which promote CO_2 electroreduction. Theory further demonstrated that CO formation was favoured because the free-energy barrier to COOH formation was decreased and the adsorption of H was impeded. This coordination environment, combined with the high surface area, made more active sites accessible during catalysis and thus promoted CO_2 electroreduction.

Lead-

PbSDB and CdSDB, where SDB is 4,4′-sulfonyldibenzoate, are structurally-related metal–organic frameworks that have potential for selective CO_2 adsorption. The structural stabilities and guest–host interactions between CO_2 and PbSDB or CdSDB frameworks at high pressures up to 13GPa *in situ* were investigated[152] using Raman spectroscopy, FTIR spectroscopy and synchrotron X-ray diffraction. Both empty frameworks exhibited high chemical stabilities under compression, but they exhibited differing pressure-induced changes in crystallinity. This difference was attributed to their differing coordination topologies, which resulted in a near-isotropic contraction of the unit cell for the CdSDB framework but anisotropic changes in the case of the PbSDB framework. The CO_2-loaded PbSDB and CdSDB frameworks at high pressures had very contrasting guest–host interactions in terms of pressure-regulated CO_2 adsorption sites. In both frameworks, pressure could very efficiently promote the formation of new CO_2 adsorption sites and the enhancement of guest–host interactions. In the CO_2-loaded PbSDB framework, in particular, the peculiar pressure-tuned CO_2 population was observed preferentially on one of the two adsorption sites in response to external compression. These unique guest–host interaction behaviors can also be unambiguously correlated with their differing topological origins. The findings for the PbSDB and CdSDB frameworks provided in-depth understanding of the structure–property relationship, and was of fundamental importance for CO_2 storage application in SDB-based metal-organic frameworks. An investigation of the structural stabilities and CO_2 adsorption behaviors of CdSDB and PbSDB at pressures of up to about 13GPa showed that both activated CdSDB and PbSDB frameworks underwent structural modification to an amorphous state, with their chemical structure intact.

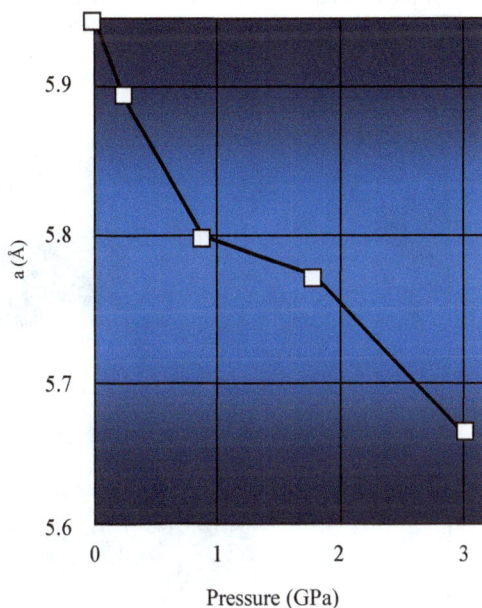

Figure 6. Unit-cell a-axis parameter of the
PbSDB framework as a function of pressure

Upon decompression, structural changes in CdSDB were completely reversed, while PbSDB exhibited reduced crystallinity. The differing compression behaviors were attributed to the near-isotropic CdSDB coordination and the non-isotropic PbSDB coordination; as reflected in the pressure dependences of the unit-cell parameters of the two frameworks. The CdSDB has two CO_2 adsorption-sites at ambient pressures, but only one spectroscopically degenerate site was detected at low pressures in CO_2-loaded CdSDB. The degeneracy disappeared at high pressures, leading to the observation of a second adsorption site. The PbSDB had only one CO_2 adsorption site at ambient pressures, but exhibited a new adsorption site under compression. The new adsorption site was structurally favoured, as indicated by CO_2 population growth under pressure until it became the only populated site at high pressures. Formation of the new CO_2 adsorption site in the PbSDB was explained in terms of a pressure-induced structural change that was associated with symmetry-breaking via phenyl-ring rotation. Variations in the pressure-tuned CO_2 populations, between the 2 sites, was attributed to the differing non-isotropic responses of the unit-cell parameters (figures 6 to 9), which led to a change in the critical distances for the formation and destruction of CO_2 adsorption sites within the framework.

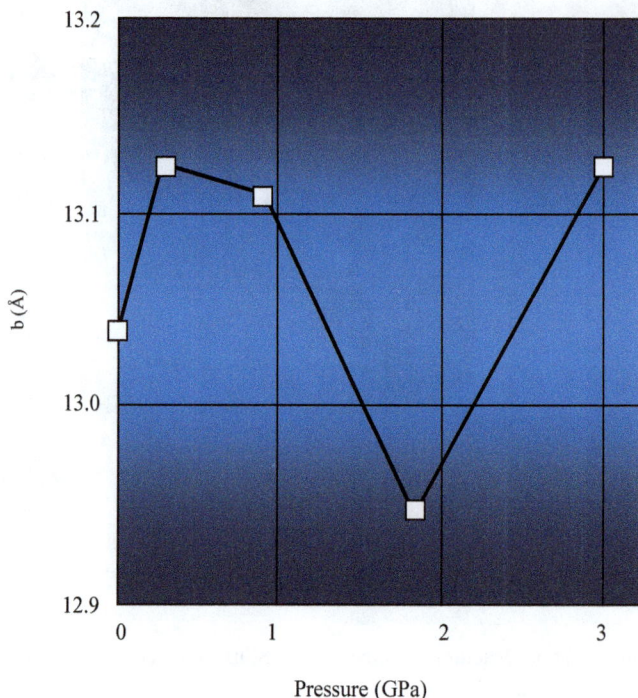

Figure 7. Unit-cell b-axis parameter of the PbSDB framework as a function of pressure

Lithium-

UTSA-16 is the second highest porous MOF for CO_2 capture, and this is attributed to its microporous structure of anatase type and the fact that the K^+ species located in the channel can interact with CO_2 molecules. A series of alkali metal cation-exchanged variants, M-UTSA-16, with M = Li, Na, K, Rb or Cs, was prepared[153] and evaluated for CO_2 capture. The CO_2 adsorption isotherms at 273K and 298K showed that the adsorption capacity for CO_2 decreased in the order: $K^+ > Na^+ > Li^+ > Rb^+ > Cs^+$. This series was used as catalysts for the transformation of CO_2 and epoxide into cyclic carbonate in the absence of a co-catalyst. The Li-UTSA-16 exhibited the highest efficiency of catalytic activity, but this was inconsistent with the sequence of CO_2 adsorption capacities. Further investigation showed that the decreasing order of catalytic activity of the M-UTSA-16 corresponded to an increasing radius of the exchanged cations and of the heat of adsorption for CO_2 at lower pressures.

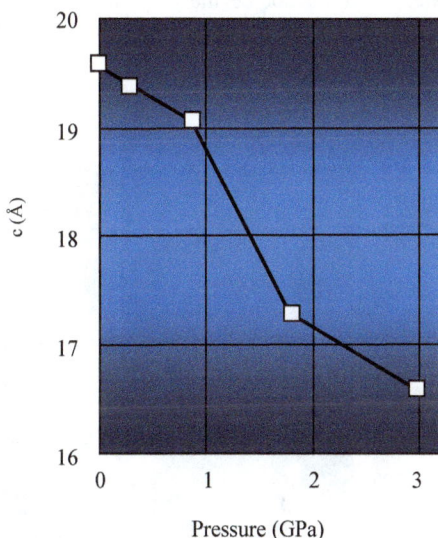

Figure 8. Unit-cell c-axis parameter of the PbSDB framework as a function of pressure

Magnesium-

The Mg_2(dobpdc), 4,4'-dihydroxy-(1,1'-biphenyl)-3,3'-dicarboxylic acid materials can be diamine-functionalized in various ways[154]: en (ethylenediamine), een (N-ethylethylenediamine), ipen (N-isopropylethylenediamine). A slight change in the number of alkyl substituents on the diamines and their alkyl chain-length can dictate the desorption temperature at 100%CO_2 and other characteristics. The presence of bulky substituents on the diamines increases framework stability upon exposure to O_2, SO_2 and H_2O vapour. Among these diamine-functionalized metal–organic frameworks, 1-een is a promising material for sorbent-based CO_2 capture. A ^{15}N solid-state nuclear magnetic resonance spectroscopic study of diamines attached to Mg_2(dobpdc) revealed[155] that both bound and free amine nitrogen environments existed when diamines were coordinated to the open Mg^{2+} sites of the framework, and rapid exchanges occurred between the 2 nitrogen environments. The activation energy required for these exchanges were a measure of the metal-amine bond-strength. A negative correlation between the metal-amine bond-strength and the CO_2 adsorption step pressure indicated however that the metal-amine bond strength could not be the only important factor affecting the CO_2 adsorption properties of diamine-functionalized Mg_2(dobpdc). It exhibited a gradual increase in CO_2 take-up, with increasing temperature, when tetraethylenepentamine was heavily loaded into its framework[156].

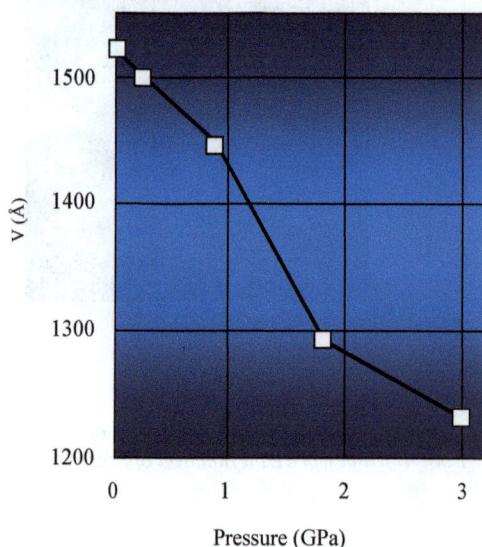

Figure 9. Unit-cell volume of compressed
PbSDB as a function of pressure

At moderate loadings, samples exhibited a working capacity of 11.2wt% and cyclable performance at 40C and 15% CO_2 pressure. A sudden increase in the CO_2 isotherms indicated that tetraethylenepentamine-loading involved a CO_2-insertion mechanism. All of the amine groups in the tetraethylenepentamine could participate to some extent in the formation of ammonium carbamate ion-pairs. A high loading of tetraethylenepentamine permitted selective CO_2 take-up ahead of H_2O vapour, and imparted an increased framework stability in humid conditions. Diamine-functionalized metal-organic frameworks which had S-shaped CO_2 isotherms were studied[157] by means of vacuum swing adsorption. Attempts to maximize CO_2 purity and recovery revealed that there was a pivotal link between feed-temperature, evacuation pressure and process performance. The frameworks which attained the target, of a CO_2 purity greater than 95% and a recovery better than 90%, were mmen-Mn_2(dobpdc) and mmen-Mg_2(dobpdc). The adsorbents, mmen-Mn_2(dobpdc) and mmen-Mg_2(dobpdc) had minimum parasitic energies of 142 and 152kWhe/tonne$_{CO2}$cap, respectively, and corresponding maximum productivities of 0.40 and 0.45mol$_{CO2}$/m^{3}adss; all with a CO_2 purity greater than 95% and a recovery better than 90%. The lower energy consumption was attributed to a perhaps low N_2 affinity and to the shape of the CO_2 isotherm. The separation of CO_2 from other components of flue gases is especially difficult due to the typically low (*circa* 40mbar) CO_2 partial pressures. Suitable materials must consequently bind strongly with CO_2 at partial pressures below 4mbar in order to capture more than 90% of the target effluent. High partial pressures of oxygen and water vapour can also present problems. Here again, functionalization of Mg_2(dobpdc) can help. Attachment of cyclic diamine 2-(aminomethyl)piperidine (2-ampd) produces[158] an adsorbent which can capture more than 90% of the CO_2 in a humid natural-gas flue emission stream. It captures the CO_2 via a cooperative mechanism which gives access to a large CO_2-cycling capacity for a small temperature-swing: 2.4mmol$_{CO2}$/g with $\Delta T = 100C$. It is suggested that water enhances CO_2-capture by 2-ampd-Mg_2(dobpdc) via hydrogen-bonding interactions with the carbamate groups of ammonium carbamate chains which form during CO_2 adsorption, thus increasing the thermodynamic driving force for CO_2 binding.

Porous coordination polymers of Mg^{II} dihyroxyterephtalate (Mg-MOF-74) with various Pt_x loadings (x = 0 to 10wt%) were synthesized and tested for catalytic cyclo-addition reaction between CO_2 and propylene oxide. The reaction was studied over a range of temperatures (100 to 170C), pressures (9.1 to 19.5bar) and reaction times (4 to 15h) to produce propylene carbonate. The results[159] showed that the platinum loading on the catalyst surface improved the selectivity toward propylene carbonate in the presence of dimethylformamide and dichloromethane, which served as solvents and promoters for the reaction. The platinum loading was related to the number of uncoordinated MgO defect sites produced during synthesis. As the platinum loading increased, the number of uncoordinated MgO sites in the Mg-MOF-74 framework structure also increased. During reaction, propylene oxide conversion and propylene carbonate selectivity increased systematically with platinum loading and the number of uncoordinated MgO defect sites.

A synergistic effect was observed, in that the combined use of Pt_x/Mg-MOF-74, dimethylformamide and dichloromethane yielded benefits beyond their individual use. A novel hybrid metal organic framework was developed by seeding the crystal growth of Mg-MOF-74 from the surface of carbon nanotube mats, so-called bucky-papers. The seeding density and growth kinetics of the metal-organic framework crystals across the carbon nanotube bucky-papers was increased by continuous discharge plasma treatment with O_2/Ar or NH_3 gas-streams. X-ray photo-electron spectroscopy was used[160] to demonstrate the impact of plasma treatment upon the anchoring of hydroxyl, amine and carbonyl functional groups to graphitic surfaces, resulting in higher surface wettability and greater metal-organic framework-seeding densities. The CO_2 adsorption isotherms at 25C showed a large increase in CO_2 take-up (maximum of 10.70mmol CO_2/g) for MOF-CNT-BP samples compared to virgin CNT-BP (0.35mmolCO_2/g) or parent Mg-MOF-74 (maximum of 3.13mmol CO_2/g). A higher CO_2 adsorption was attributed to the synergistic effect of metal-organic framework and carbon nanotube, due to an increased porosity at the interface, to better metal-organic framework crystal distribution and to increased dispersive forces which improved the structural integrity of the framework component.

Manganese-

A new 3-dimensional porous Mn^{II}-based metal-organic framework $[Mn_4(L)_2(H_2O)_4]_n \cdot 4DMF \cdot H_2O$ (1) has been synthesized[161] using Mn^{II} and the rigid 2,6-di(2′,5′-dicarboxylphenyl)pyridine (H4L) ligand under solvothermal conditions. Structural analysis revealed that MOF 1 was an open framework with exposed active nitrogen atoms and rectangular-section channels along the b-axis, with effective aperture sizes of 11.3Å x 8.8Å and 10.7Å x 8.7Å. Gas sorption behaviors showed that MOF 1 has a relatively high capacity and selectivity for CO_2 over CH_4. Thus CO_2 cyclo-addition was studied using various epoxides in order to demonstrate the high efficiency of the catalyst.

Two new isostructural microporous coordination frameworks $[Mn_3(Hpdc)_2(pdc)_2]$ and $[Mg_3(Hpdc)_2(pdc)_2]$ (pdc2- = pyridine-2,4-dicarboxylate) showing primitive cubic (pcu) topology have been prepared[162]. The pore aperture of the channels was too narrow for the efficient adsorption of N_2; however, both compounds demonstrated a substantially higher take-up of CO_2 (119.9ml/g for $[Mn_3(Hpdc)_2(pdc)_2]$ and 102.5ml/g for $[Mg_3(Hpdc)_2(pdc)_2]$ at 195K, 1bar). Despite their structural similarities, $[Mg_3(Hpdc)_2(pdc)_2]$ had a typical reversible type-I isotherm for adsorption/desorption of CO_2, while $[Mn_3(Hpdc)_2(pdc)_2]$ featured a 2-step adsorption process with a very broad hysteresis between the adsorption and desorption curves.

Two novel metal-organic frameworks, [Mn(CIP-)2] (1) and [Ag(CIP-)] (2) (HCIP = 4-(4-carboxylphenyl)-2,6-di(4-imidazol-1-yl)phenyl)pyridine), were solvothermally synthesized[163], based on a pyridyl-imidazole-carboxyl multifunctional ligand. Single-crystal X-ray diffraction analysis showed that complex-1 is a 3-dimensional microporous framework with uncoordinated imidazole groups, and that complex-2 is a 2D + 2D → 2D 3-fold parallel interpenetrated network. Complex-1 exhibited excellent CO_2 selective absorption over N_2 and CH_4. Calculations revealed that the selectivities of complex-1 for the CO_2/CH_4 (50:50) and CO_2/N_2 (15:85) mixtures were 8.0 and 117 at 273K under 1bar, respectively. Moreover, luminescence investigations showed that complex-2 is an excellent MOF-based multiresponsive fluorescent probe for Fe^{3+}, $CrO_4^{2-}/Cr_2O_7^{2-}$ and 2,6-dich-4-nitroaniline, with high selectivity and sensitivity. Complex-2 exhibited a highly sensitive sensing ability (5.2 x 10^4/M) and a low detection limit (1.7 x 10^{-7}M) for 2,6-dich-4-nitroaniline.

A microporous 3-dimensional supramolecular Mn^{II}-porphyrin framework material, [{Mn(TCPP)$_{0.5}$(H$_2$O)$_2$}•2H$_2$O]$_n$, where TCPP is 5,10,15,20-tetrakis(4-carboxyphenyl) porphyrin, exhibited[164] a striking visible-light assisted cyclo-addition of CO_2 to epoxides to create carbonates at room temperature and 1bar. The structure had a microporous 2-dimensional network structure, with 1-dimensional 6.32Å x 11.88Å channels along the c-axis and with 2 types of 1-dimensional channel, 3.94Å x 8.37Å and 4.66Å x 4.93Å, along the a-axis[165]. The 2-dimensional networks also interacted via O-H⋯O hydrogen-bonding between carboxylate oxygen and the hydrogen of water molecules which were coordinated to Mn^{II} centers, thus forming a 3-dimensional supramolecular framework. Its gas-adsorption selectivity constants were 39 for CO_2/H_2, 19 for CO_2/N_2 and 21 for CO_2/Ar; with an isosteric heat-of-adsorption of 32.9kJ/mol.

Molybdenum-

A novel polyoxometalate (POM)-based metal-organic framework, TBA$_5$[P$_2$Mo$_{16}$VMo$_8$ VIO$_{71}$(OH)$_9$Zn$_8$(L)$_4$] (NNU-29), was *in situ* synthesized[166] and applied to CO_2 photoreduction. The selection of porous material containing a reductive POM cluster was considered to be helpful for CO_2 reduction; meanwhile, a hydrophobic-group-modified organic ligand permitted NNU-29 to exhibit a good chemical stability and to restrict hydrogen generation to some extent. In photocatalytic CO_2 reduction, the yield of HCOO- reached 35.2μmol in aqueous solution, with a selectivity of 97.9% after 16h.

Nickel-

Two novel metal-organic frameworks [Ni$_2$(μ_2-Cl)(BTBA)$_2$ DMF]·Cl3DMF (JLU-MOF56, BTBA = 3,5-bis(triazol-1-yl)benzoic acid) and [Co$_2$(μ_2-Cl)(BTBA)$_2$·DMF]·Cl·3DMF (JLU-MOF57) have been synthesized[167] under solvothermal conditions. The 2 compounds are isoreticular and are constructed from binuclear [M$_2$(μ_2-Cl)(COO)$_2$N$_4$] (M = Co, Ni) and a 3-connected hetero-N,O donor ligand. The overall framework possesses a (3,6)-connected dag topology, and both contain ultramicroporous channels of 3.5Å x 3.4Å, which are suitable for adsorbing carbon dioxide molecules but not the larger nitrogen molecules.

A new 3-dimensional framework, [Ni(btzip)(H$_2$btzip)]$_2$ DMF·2H$_2$O (H$_2$btzip=4,6-bis(triazol-1-yl)isophthalic acid), for use as an acidic heterogeneous catalyst was constructed by the reaction of nickel wire and a triazolyl-carboxyl linker[168]. The framework consisted of intersecting 2-dimensional channels decorated with free COOH groups and uncoordinated triazolyl N atoms, leading not only to a high CO$_2$ and C$_2$H$_6$ adsorption capacity but also to significant selective capture for CO$_2$ and C$_2$H$_6$ over CH$_4$ and CO at 273 to 333K. Moreover, the material possessed chemical stability towards water. Grand canonical Monte Carlo simulations confirmed the existence of multiple CO$_2$- and C$_2$H$_6$-philic sites. As a result of the presence of accessible Brønsted acidic COOH groups in the channels, the activated framework demonstrates highly efficient catalytic activity in the cyclo-addition reaction of CO$_2$ with propylene oxide/4-chloromethyl-1,3-dioxolan-2-one/3-butoxy-1,2-epoxypropane to give cyclic carbonates.

Isomorphic (Ni$_2$(l-asp)$_2$bipy and Ni$_2$(l-asp)$_2$pz) with different pore sizes were incorporated into poly(ether-block-amide) (Pebax-1657) to prepare mixed-matrix membranes with gas-permeation properties for CO$_2$, H$_2$, N$_2$ and CH$_4$. Different loading ratios of MOFs with mass percentages ranging from 10% to 30% were studied[169]. Compared with the pure polymer membrane, the two as-synthesized series of MMMs showed an improved CO$_2$ permeability and CO$_2$/H$_2$ selectivity. The highest CO$_2$ permeation property was achieved by Ni$_2$(l-asp)$_2$bipy-Pebax-20 of 120.2 barrer with an enhanced CO$_2$/H$_2$ selectivity of 32.88 compared with the 55.85 barrer and 1.729 of the pure polymer membrane, respectively, which makes it a potential candidate for future applications in CO$_2$ capturing.

A new MOF/GO composite based upon [Ni$_2$(BDC-NH$_2$)$_2$(DABCO)].xDMF.yH$_2$O metal-organic framework (abbreviated as Ni-A; BDC-NH$_2$ = 2-aminoterephthalic acid; DABCO = 1,4-diazabicyclo[2.2.2]octane)) and graphene oxide has been used[170] for CO$_2$ and N$_2$ adsorption at ambient pressure. The Ni-A/GO nanocomposites featured Ni-A crystals grown on the graphene surface along the [002] direction. The obtained composites exhibited various morphologies, such as nanosheets and rectangular shapes, as compared to the parent Ni-A. The adsorption of CO_2 and N_2 was individually assessed and their isotherms were fitted by using the single-site Langmuir model. Selectivities of the samples for CO_2/N_2 mixtures in the ratio of (0.15/0.85) at 273K were estimated using the ideal adsorbed solution theory (IAST). Although the CO_2 take-ups of the composites were quite similar or somewhat lower as compared to Ni-A, the selectivity for CO_2/N_2 adsorption on all of the composites was found to be higher. Among the synthesized composites with differing GO contents, Ni-AG5 (with 5% GO) exhibited the maximum CO_2/N_2 selectivity (52 at 273K and 1bar), which is 3 times greater than the selectivity of the parent MOF under the same conditions. However, gas take-up by the Ni-AG5 composite was similar to that of the parent MOF (6.8mmol/g at 273K and 1bar).

A cheap and commercially available formic acid ligand was used[171] to construct a robust MOF material [$Ni_3(HCOO)_6 \cdot DMF$], offering high chemical stability, low cost and high selectivity towards C_2H_2 over CO_2. The exceptional separation performance of the activated material is attributed mainly to the small pore size (4.3Å) and functional O donor sites on the pore walls that provide a strong binding affinity towards C_2H_2, as revealed by detailed computational studies. This material thus exhibits ultrahigh low-pressure C_2H_2 take-up (38.2cm^3/cm^3 at 0.01bar and 298K) and possesses a high C_2H_2/CO_2 selectivity (22.0 at ambient conditions), comparable to that of other leading porous materials. A high separation performance was further confirmed by breakthrough experiments performed on a 50/50 C_2H_2/CO_2 mixture.

A thermostable anion-pillared metal-organic framework TIFSIX-2-Ni-i (also known as ZU-12-Ni, TIFSIX = hexafluorotitanate, 2 = 4,4'-bipyridylacetylene, i = interpenetrated) was synthesized for the first time. Structural characterizations showed that TIFSIX-2-Ni-i exhibits a 2-fold interpenetrated network with a narrow aperture of 5.1Å (H-H distance) and abundant inorganic anion sites in the channel[172]. The subtle change in metal and inorganic linkers results in the enhancement of the thermal stability of TIFSIX-2-Ni-i. The TIFSIX-2-Ni-i exhibited high separation abilities, especially for the important separations of C_2H_2/C_2H_4 and C_2H_2/CO_2 mixtures. Adsorption data demonstrate that TIFSIX-2-Ni-i enables effective C_2H_2 capture (4.21 mmol/g, at 298K and 1bar), and achieves highly selective separation of C_2H_2 over C_2H_4 or CO_2 with an IAST selectivity of 22.7 (C_2H_2/C_2H_4: 1/99) and 10.0 (C_2H_2/CO_2: 2/1) at 298K and 1bar. Dispersion-corrected density functional theory (DFT-D) calculations confirm the preferential adsorption of C_2H_2 over C_2H_4 and CO_2. Furthermore, the potential of industrial feasibility of TIFSIX-2-Ni-i for C_2H_2/C_2H_4 separation is confirmed by transient breakthrough tests (611.4mmol/L C_2H_2 absorbed, for mixtures containing 1% C_2H_2).

The integration of additional functionality in an amine-decorated metal-organic framework by encapsulating nickel nanoparticles was reported[173]. In-depth characterization of the post-modified structure confirmed the presence of well-dispersed and ultra-small nanoparticles within the framework pores. Although the surface area was more reduced than pristine MOF, the CO_2 take-up capacity was markedly increased by 35%, with a large 10kJ/mol rise in adsorption enthalpy which reflected favourable interactions between the CO_2 and nanoparticles. In particular, CO_2 adsorption selectivity over N_2 and CH_4 displays significant improvement (CO_2/N_2 = 145.7, CO_2/CH_4 = 12.65), while multi-cycle CO_2 take-up demonstrates outstanding sorption recurrence. Impressively, the embedded NPs act as highly active functional sites toward solvent-free CO_2 cyclo-addition with epoxides in 98% yield and 99% selectivity under relatively mild conditions. The catalyst shows high recyclability without leaching of any metal-ion/nanoparticles, and a greater pre-eminent activity than that of unmodified analogues. The outstanding conversion and selectivity properties were maintained for a wide range of aliphatic and aromatic epoxides. Larger substrates exhibited insignificant conversion, thus demonstrating considerable size-selectivity.

Two isostructural Ni^{II} redox-active metal–organic frameworks containing flexible tripodal trispyridyl ligand and linear dicarboxylates such as terephthalate and 2-aminoterephthalate were studied mainly for their use in oxygen evolution. The 2-dimensional-layered metal-organic frameworks formed 3-dimensional hydrogen-bonded shapes containing 1-dimensional hydrophilic channels that were filled with water molecules. As well as aiding in oxygen evolution, the materials exhibited[174] a marked ability to absorb water vapor (180 to 230cc/g at 273K) and CO_2 (33cc/g at 273K).

Aqueous synthesis of metal-organic framework materials having nickel and cobalt as the metals, and benzene-1,4-dicarboxylic acid as the linker, could be carried[175] out rapidly in high yields by using microwave irradiation. The resultant compound could induce high conversion-rates of epoxides to cyclic carbonates, with greater-than 99% selectivity, under solvent-free conditions. The bimetallic material exhibited a greater catalytic activity than that of the corresponding single-metal catalysts. This was attributed to a synergistic catalytic effect arising from charge transfer between the nickel and cobalt metal centers.

A highly porous 3-dimensional framework, $[Ni_3(BTC)_2(MA)(H_2O)](DMF)_7$, which incorporated exposed metal sites and nitrogen-rich melamine was prepared[176] via the solvothermal assembly of H_3BTC (1,3,5-benzenetricarboxylic acid), MA (melamine) and Ni^{II} ions. The resultant material had a high CO_2 loading capacity and CO_2 affinity, due to Lewis-base properties and a high micropore density. The catalytic activity was such as to ensure high-efficiency CO_2 cyclo-addition reaction with epoxides under ambient conditions. The material could also be recycled with no obvious decrease in catalytic activity.

A new microporous Ni-based metal-organic framework (Ni-MOF-1) was synthesized[177] by the reaction of $Ni(NO)_3 \bullet 6H_2O$ with the binary ligands of tetracarboxylic 2,3,5,6-tetrakis(4-carboxyphenyl)pyrazine (H_4TCPP) and 4,4′-bipyridine under solvothermal reaction conditions. Single-crystal X-ray diffraction data indicated that Ni-MOF-1 had a 3-dimensional porous network with novel trinuclear $Ni_3(OH)_2(COO)_4(HCOO)_4$ clusters, and exhibited CO_2 take-ups of 37.57cm^3/g and a selectivity of CO_2 over N_2 of 42.89 at 273K.

An open metal site framework named, UTSA-16, was synthesized[178] and modified as a high-capacity adsorbent for reversible CO_2 capture. Partial substitution of intrinsic Co^{2+} sites of UTSA-16 with Ni^{2+} centres was achieved over the molar composition range of 0 to 75%Ni with the aim of increasing CO_2 take-up. Synthesized bimetallic Ni_x-UTSA-16 (x = 0, 20, 50, 75) materials were characterized by using various techniques to assess the influence of chemical composition upon CO_2 binding affinity and any subsequent physical change in morphology, crystal size and porosity upon the total take-up. Experimental isotherm adsorption studies showed the following trend for CO_2 adsorption capacity employing the Ni_x-UTSA-16 series: Ni20-UTSA-16 > UTSA-16 > Ni50-UTSA-16 > Ni75-UTSA-16. According to the dynamic breakthrough CO_2 profiles measured for a mixture of CO_2 and CH_4 (15/85 molar ratio), Ni20-UTSA-16 exhibited twice the breakthrough time with 1.5 times the loading capacity at 75Nml/min feed flow rate, as compared to the parent UTSA-16. In addition, the Ni20-UTSA-16 bimetallic metal–organic framework exhibited a lower isosteric heat of adsorption as compared to UTSA-16 (ΔH_{ave} = 28.54 versus 46.85kJ/mol). As a result, more than 95% of its capacity was restored by applying a partial vacuum for only 1h at room temperature without involving any other time- or energy-consuming regenerative step.

Ruthenium-

A series of efficient heterogenized ruthenium catalysts, Ru_x-NHC, (x = 1, 2, 3; NHC = N-heterocyclic carbene) were synthesized[179] by the immobilization of various ruthenium complexes $RuCl_3$, $[RuCpCl_2]_2$ (Cp = pentamethylcyclopentadienyl) and $[Ru(C_6Me_6)Cl_2]_2$ (C_6Me_6 = hexamethylbenzene) on an azolium-based metal-organic framework via post-synthesis metalation. The heterogenized catalysts were used for the catalytic hydrogenation of CO_2 to formic acid. The Ru_3-NHC catalyst exhibited the highest activity because of the strong electron-donating nature of the C_6Me_6 ligand of the $[Ru(C_6Me_6)Cl_2]_2$ complex, which is favourable for CO_2 hydrogenation. A turnover number of up to 3803 was obtained at 120C under a total pressure of 8MPa (H_2/CO_2 = 1) within 2h with K_2CO_3 additive in N,N-dimethylformamide solvent. The heterogenized Ru_3-NHC catalyst could be recovered by filtration, with little loss of catalytic activity.

Metal-organic frameworks, ZIF-67, with different morphologies were synthesized[180] via a solvent-induced method at room temperature. The photocatalytic performance with regard to the reduction of CO_2 was evaluated by using ZIF-67 materials as co-catalysts, cooperating with a ruthenium-based complex as the photosensitizer. It has been demonstrated that the 2-dimensional ZIF-67 with a leaf-like morphology exhibited the best photocatalytic activity and stability due to having the highest CO_2 adsorption capability and efficient electron transfer from the excited $[Ru(bpy)_3]_2+$ to ZIF-67.

Samarium-

Highly thermal and chemically stable, 20-connected lanthanide metal-organic frameworks, $[\{Ln(BTB)(H_2O)\}\cdot H_2O]_n$ (where Ln = Sm (MOF1)/Gd (MOF2), BTB = 1,3,5-tris (4-carboxy phenyl) benzene) have been synthesized[181] solvothermally, and characterized by single-crystal X-ray diffraction analysis and other physicochemical methods. MOF1 and 2 are isostructural and feature three-dimensional honeycomb-like structures with large one-dimensional hexagonal channels with dimensions of ~10.20Å x 10.11Å. Gas take-up studies of the samples revealed selective adsorption properties of MOF1 for CO_2 over other (N_2, Ar, H_2) gases. The activated samples of the MOF1/2 act as efficient recyclable catalysts for the heterogeneous cyclo-addition of CO_2 with styrene oxide, resulting in cyclic carbonate with high yield and selectivity. Interestingly, a pore size-dependent catalytic conversion of epoxides has been observed, suggesting the potential utility of MOF1 as a promising heterogeneous catalyst for the cyclo-addition of carbon dioxide. Furthermore, the MOF1 catalyst can be easily recycled for several cycles without significant loss of catalytic activity or structural rigidity.

Scandium-

A new amide-functionalized metal-organic framework, $[Sc_3(\mu_3-O)(L)_{1.5}(H_2O)_3Cl]_n$ [NJU-Bai49; H4L = 5-(3,5-dicarboxybenzamido)isophthalic acid], having the combined features of highly selective CO_2 adsorption and high thermal and chemical stability was synthesized[182] and exhibited a 4.5wt%CO_2 take-up at 298K and 0.15bar, plus the highest (166.7) selectivity for CO_2/N_2 to be observed among all amide-functionalized metal-organic frameworks. Monte Carlo simulations further indicated that both the decorated amide group and the open metal site acted as CO_2 binding sites, and probably contributed to its highly selective CO_2 take-up.

Silver-

A non-interpenetrating organosulfonate-based metal-organic framework with a defective primitive-cubic (pcu) topology was synthesized[183]. The unusual missing linkers, along with the highest permanent porosity (~43%) in sulfonate-MOFs, offered a versatile platform for the incorporation of alkynophilic Ag^I sites. The cyclic carboxylation of alkyne molecules (e.g., propargyl alcohol and propargyl amine) into α-alkylidene cyclic carbonates and oxazolidinones were successfully catalyzed by the use of Ag^I-embedded sulfonate-MOF under an atmospheric pressure of CO_2. In all of the three catalytic reactions using CO_2 as feedstock, the highly robust sulfonate-based MOF catalyst gave at least 3 cycles of re-usability.

A novel 3-dimensional cationic metal–organic framework, $\{[AgTPB]\cdot SbF_6\}_n$, based upon AgI and 1,2,4,5-tetra(4-pyridyl) benzene was prepared[184] using the diffusion synthesis method. The counter anion SbF_6^- coming from $AgSbF_6$ existed in the pores. The activated material could selectively adsorb CO_2 over CH_4 and H_2O over C_2H_5OH.

Table 8. Self-Diffusion of $^{13}CO_2$ in $Zn_2(dobpdc)$ at 298K

Pressure (mbar)	$D_{\parallel}$ (m²/s)	$D_{\perp}$ (m²/s)	$D_{\parallel}/D_{\perp}$
635	5.8×10^{-9}	1.9×10^{-10}	30
1010	6.2×10^{-9}	2.3×10^{-10}	27
2026	6.5×10^{-9}	1.4×10^{-10}	48

Terbium-

By employing a tricarboxylate ligand 1-(4-carboxybenzyl)-1H-pyrazole-3,5-dicarboxylic acid (H_3L), four lanthanide metal-organic frameworks (Ln-MOFs) formulated as $\{[LnL(H_2O)_2]\cdot H_2O\}_n$ (Ln = Eu, Gd, Tb, Dy) have been prepared[185] under hydrothermal conditions. Single-crystal X-ray analyses revealed that the 4 compounds were isomorphous and exhibited a 3-dimensional network structure featuring a 1-dimensional rectangular channel with a size of about 7.8Å x 12.1Å along the b-axis. The frameworks had a marked stability at high temperatures, in humid air, in water, as well as in acid/base environments. Efficient ligand-sensitized characteristic luminescence was observed in the visible region for europium- and terbium-based compounds. The terbium material was a multifunctional material which could quantitatively detect Fe^{III} ions in Fe^{II}/Fe^{III} solutions, selectively adsorb CO_2 over CH_4, and be used as a catalyst in cyclization reactions with epoxides and CO_2.

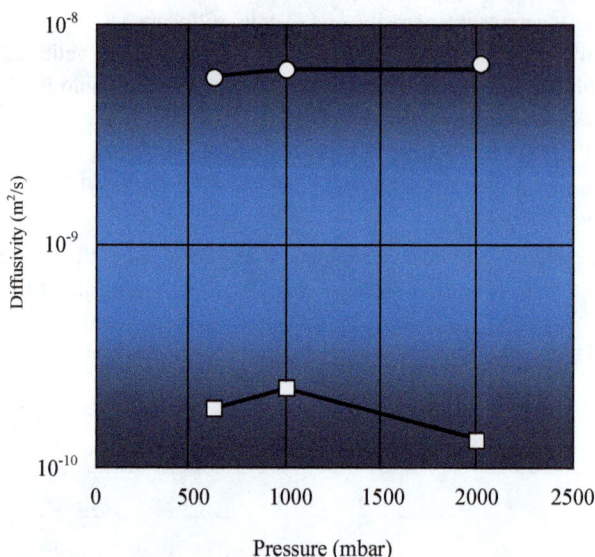

Figure 10. Self-diffusion of $^{13}CO_2$ in Zn$_2$(dobpdc) crystals at 25C circles: parallel to c-axis, squares: perpendicular to c-axis

Titanium-

Nanosized semiconductors with hierarchical structures are attractive for use as catalysts. By using a well-defined titanium-based metal organic framework, i.e., MIL-125(Ti), as a template, the synthesis of Ni-doped mesoporous TiO_2 nanocrystals with tablet morphologies can be carried out. While the obtained Ni-TiO_2 composites could be used[186] for the photoreduction of CO_2 in the presence of water vapor, this photocatalytic property could be greatly improved by the further deposition of silver nanoparticles. In principle, nickel species in the lattice matrix led to the formation of impurity levels in the band-gap of TiO_2, thus promoting both light absorption and charge separation. On the other hand, silver as a co-catalyst could trap electrons and simultaneously activate the C-O bonds from the adsorbed CO_2. The results demonstrated that 1.0%Ag/0.5%Ni-TiO_2

photocatalyst exhibited the highest activity. The yields of CO and CH_4 attained 14.31 and 279.07mol-g, respectively, during 3h of reaction.

Zinc-

By introducing a 1,2,4-triazole ligand into the tetracarboxylate system, an anionic metal-organic framework $[(CH_3)_2NH_2][Zn_2(ABTC)(Tz)] \bullet 3DMF$, where ABTC is 2,2',5,5'-azobenzene tetracarboxylic acid, Tz is 1,2,4-triazole and DMF is N,N-dimethylformamide, was synthesized[187] by using solvothermal methods. The compound exhibited a CO_2 adsorption capacity of 92.1cm^3/g at 273K under 1atm., a good adsorption capacity for small molecules such as CH_4, C_2H_6 and C_3H_8 and an excellent separation selectivity for CO_2/CH_4, C_2H_6/CH_4 and C_3H_8/CH_4.

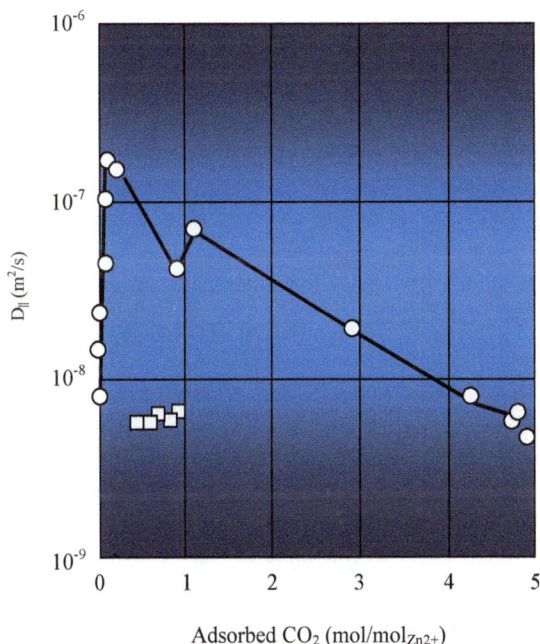

Figure 11. Self-diffusivity of CO_2 at 298K along the c-axis
circles: molecular dynamics rigid-lattice model predictions,
squares: pulsed field gradient nuclear magnetic resonance data

A detailed investigation has been made[188] of CO_2 diffusion within the pores of Zn_2(dobpdc), where dobpdc^{4-} is 4,4'-dioxidobiphenyl-3,3'-dicarboxylate, by means of pulsed field gradient nuclear magnetic resonance spectroscopic and molecular dynamics techniques. This permitted precise determination of the self-diffusion coefficients. Observation of CO_2 diffusion through channels parallel to the c-axis indicated a self-diffusion coefficient, $D_{\parallel}$, of $5.8 \times 10^{-9} m^2/s$ at a CO_2 pressure of 625mbar. Measurements which were performed at pressures of between 625 and 2026mbar indicated very similar diffusion values (table 8, figure 10), whereas molecular dynamics simulations predicted large variations in self-diffusion at higher pressures (figure 11). Such observations incidentally revealed that the CO_2 could also diffuse between the hexagonal channels in the crystallographic ab-plane with a diffusivity, $D_{\perp}$, of $1.9 \times 10^{-10} m^2/s$ even though these channel walls appeared to be impermeable according to single-crystal X-ray crystallographic and flexible-lattice molecular dynamics simulations. This diffusivity in the ab-plane hinted at the existence of defects that could permit effective multi-dimensional CO_2 transport to occur in metal-organic frameworks which ostensibly possessed only 1-dimensional porosity.

Based upon a structural analogy between zinc coordination in metal-organic frameworks and that in carbonic anhydrase, it has been realized[189] that the former can mimic the latter and thus convert CO_2 into other products. Carbonic anhydrase is a zinc-containing metalloprotein in which the zinc active center plays a key role in transforming CO_2 into carbonate.

$$CO_2 + OH^- \Rightarrow HCO_3^-$$
$$CO_2 + OH^- \Rightarrow HCO_3^-$$
$$HCO_3^- + CaCl_2 \Rightarrow CaCO_3$$

with the first reaction being moderated by carbonic anhydrase and the second reaction being moderated by the metal-organic framework. The biomimetic activity of metal-organic frameworks was explored by detecting the hydrolysis of para-nitrophenyl acetate: a model reaction which is used to determine carbonic anhydrase activity. Metal-organic frameworks such as ZIF-100 (figure 12) were used to mimic the ability of carbonic anhydrase to convert CO_2 gas. The biomimetic conversion of CO_2 was typically carried out by first allowing 50mg of the metal-organic framework to absorb CO_2 at 1atm for 2h. The framework material was then added to 5ml of buffer solution (50mM, pH8) in order to initiate the biomimetic CO_2 reaction which converted CO_2 into HCO_3.

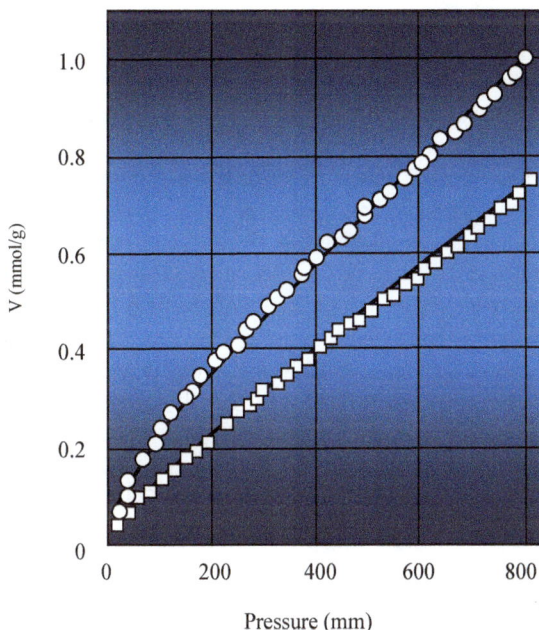

Figure 12. CO_2 adsorption take-up isotherms of ZIF-100
circles: 273K, squares: 298K

In similarly bio-inspired work, a zinc benzotriazolate metal-organic framework, $[Zn(ZnO_2CCH_3)_4(bibta)_3]$, where $bibta^{2-}$ is 5,5′-bibenzotriazolate, was subjected[190] to mild $CH_3CO_2^-/HCO_3^-$ ligand-exchange followed by thermal activation to generate nucleophilic Zn-OH groups which resembled the active site of α-carbonic anhydrase. The modified framework material, $[Zn(ZnOH)_4(bibta)_3]$, exhibited an excellent ability to capture trace CO_2 and could be regenerated at moderate temperatures. It was deduced that inter-cluster hydrogen-bonding interactions supplemented a $Zn-OH/Zn-O_2COH$ fixation mechanism. Again inspired by the structure of carbonic anhydrase, an ultramicroporous lanthanide metal-organic framework, NKMOF-3-Ln, was developed[191] which possessed pockets that could selectively bind CO_2 under ambient conditions. In particular, CO_2 molecules could be precisely located in the monocrystalline structure. Highly-ordered CO_2 molecules could interact strongly with the framework via the electrostatic interaction of nitrates, and the strong binding led to excellent CO_2 separation. The CO_2 adsorption capacity and binding energy gradually increased with lanthanide contraction. The biomimetic behavior of zinc hydroxide within a metal-organic framework having the structural and reactivity characteristics of carbonic anhydrase was studied[192]. As in the biological situation, the framework bound CO_2 molecules via insertion into the Zn–OH bond; leading to significant (3.41mmol/g) CO_2 adsorption. Like the real enzyme, the material catalyzed an oxygen-isotope exchange between water and carbon dioxide. The nodes of metal-organic frameworks are excellent mimics of metalloenzymes, due mainly to their site isolation and similar ligand fields. The framework material MFU-4l, possessing a metal node with N_3ZnX coordination, closely mimics the active site of carbonic anhydrase, with regard to structure and reactivity. The immobilized carbonic anhydrase unfortunately tends to possess a low CO_2 sequestration efficiency. A combination of carbonic anhydrase and ZIF-8 having a cruciate flower-like morphology was created[193] by adsorbing the anhydrase onto the ZIF-8. The immobilization efficiency was greater than 95%, while the maximum activity-recovery attained 75%. The shape of ZIF-8 could be controlled by adjusting the concentration of Zn^{2+} ions using high (1M) concentrations of 2-methylimidazole. The yields of $CaCO_3$ obtained by using the composites were 22 times greater than those for plain carbonic anhydrase.

Going beyond mere inspiration, the use of CO_2 and its precipitation as $CaCO_3$ was studied[194] by using immobilized bovine carbonic anhydrase. The selection of carriers exhibiting gas adsorption could increase the CO_2 sequestration efficiency of carbonic anhydrase. A metal-organic framework material, ZIF-8, was used and bovine carbonic anhydrase was encapsulated into the microporous zeolite imidazolate framework by using the so-called ship-in-a-bottle technique. Entrapment of bovine carbonic anhydrase molecules, with an enzyme loading of about 100mg/g, was carried out during the crystal growth of ZIF-8. The biocatalyst exhibited excellent behavior over wide pH and temperature ranges and was stable during storage for up to 37 days.

Interest in this topic is not limited to carbonic anhydrase. The efficient fixation of CO_2 from the atmosphere to produce useful products is another form of carbon-capture, and the coupling of enzymatic reactions with electrochemical regeneration is a promising technique. A bio-electrocatalytic system has been created[195] by depositing crystallites of the mesoporous metal-organic framework NU-1006, which contains formate dehydrogenase, onto a fluorine-doped tin oxide glass electrode modified with Rh(2,2'-bipyridyl-5,5'-dicarboxylic acid)Cl$_2$ complex. This system converts CO_2 into formic acid at the rate of 79mm/h, with electrochemical regeneration of the nicotinamide adenine dinucleotide co-factor. In tandem catalysis, different catalytic processes are arranged so as to feed the product of one reaction to the next; a method which is analogous to enzyme cascades. The method can be used[196] to up-grade small molecules such as CO_2 into more useful hydrocarbons. Metal-organic framework thin films, grown onto nanostructured gold micro-electrodes, are useful for tandem catalysis. Metal-organic framework thin films are able to suppress sharply the production of CO on nanostructured gold micro-electrodes and to produce CH_4 and C_2H_4. Enzymatic reduction of CO_2 involves a genuine multi-enzyme cascade system and, for example, formate dehydrogenase, glutamate dehydrogenase, reduced pyridine nucleotide, formaldehyde dehydrogenase and alcohol dehydrogenase have been incorporated into ZIF-8 so as to prepare 3 kinds of enzyme and co-enzyme/ZIF-8 nanocomposites[197]. These nanocomposites were then sequentially placed in a microporous membrane which was combined with a pervaporation membrane in order to separate-out the useful product, methanol. The incorporation of a pervaporation membrane helped to control the reaction direction, and the generation of methanol increased from 5.8 to 6.7µmol. The reaction-efficiency of an immobilized enzyme-ordered distribution in a membrane was higher than that of a disordered distribution, and the methanol output increased from 6.7 to 12.6µmol. The introduction of reduced pyridine nucleotide into ZIF-8 also seemed to improve the transformation, of CO_2 into methanol, from 12.6 to 13.4µmol.

The possible tuning of the CO_2-capture capability of the mixed-metal organic frameworks, mmen-$(Zn_xMg_{1-x})_2$(dobpdc), was explored[198] by using density functional theory. The stability of mixed-metal structures with respect to their parent single-metal species was deduced from the energy of mixing; the latter values were negative, thus indicating exothermic mixing and the fact that mixed-metal structures were energetically preferred to their non-interacting single-metal variants. The energy-of-mixing values of all of the mixed-metal structures were less than 1kJ/mol, thus implying that the mixing was ideal. A density-of-states analysis also showed that the electronic structures of magnesium and zinc in mixed-metal structures were not appreciably different to those of the parent single-metal species. This suggested the occurrence of weak chemical interactions between magnesium and zinc in mixed-metal structures. In CO_2-adsorbed mmen-$(Zn_xMg_{1-x})_2$(dobpdc), the adsorption energy was a linear function of x, and did not depend upon how the magnesium and zinc atoms were arranged in the various mixed-metal structures. It could be described in terms of weak chemical interactions between the magnesium and zinc in mixed-metal structures. The adsorption energy could thus be deduced from the mixing ratio.

Four new zinc porous coordination polymers were synthesized[199] from a linear ligand of [1,1'-biphenyl]-3,3',5,5'-tetracarboxylic acid (H_4L) and zinc nitrate by the strategy of high throughput solvothermal syntheses. Their structural differences arise from different pH of the medium, as well as the composition of the solvent system. Very interesting is the fact that although NTU-34, -35, and -37 possess the same 3,3,4-connections, they have completely different porous frameworks, together with a dissimilar cavity size and shape. In addition, given their same framework formula [Zn_2L], these three porous coordination polymers can be considered to be a new group of framework isomers. Furthermore, the coordination between distorted Zn_4O clusters and linear ligands with partial intermolecular $\pi\cdots\pi$ interaction formed the other porous framework (NTU-36), which has a new topology with a 3,3,3,6-connection and a point symbol of {4.62}4{42.5.64.88}2{5 x 102}{52.8}). More importantly, gas adsorption and breakthrough experiments revealed that the microporous nature enabled NTU-36 to remove CO_2 selectively from its CH_4 or N_2 mixtures under flowing conditions at ambient temperature.

A new family of quasi-3-dimensional and 3-dimensional porous coordination polymers, NTU-43 to NTU-50, was prepared[200] from a layered structure, NTU-42. Single-gas adsorption isotherms of CO_2, N_2 and CH_4 indicated the dependence of gas capacity upon pore-size, functionality and framework charge of the quasi-3-dimensional porous coordination polymers. Here, NTU-45 and NTU-46, with NH_2-BDC and OH-BDC bidentate linkers (NH_2-BDC = 2-aminoterephthalic acid and OH-BDC = 2-hydroxyterephthalic acid) exhibited a marked tendency to selective CO_2 take-up. Efficient CO_2 capture from CO_2/CH_4 and CO_2/N_2 mixtures was also confirmed by breakthrough results obtained under continuous and dynamic conditions at 298K.

A strategy was proposed[201] for the rapid *in situ* fabrication of zeolitic imidazolate framework-8 (ZIF-8) hybrid membrane with assistance of the 2-dimensional graphitic carbon nitride (g-C_3N_4) nanosheets at room temperature. The negatively charged g-C_3N_4 nanosheets with abundant nitrogen coordinating sites were able to capture and anchor Zn^{2+} and thus provide a large number of heterogeneous nucleation sites for the formation of initial ZIF-8 crystals. During cyclical spin coating of the Zn^{2+}/g-C_3N_4 nanosheets and the ligand (2-methylimidazole), the *in situ* formed ZIF-8 crystal nucleus on g-C_3N_4 nanosheets promotes the further growth of continuous defect-free membranes. The ZIF-8/g-C_3N_4 membrane with a thickness of about 240nm can be obtained by using this strategy within 0.5h at room temperature. Moreover, it exhibits promising H_2/CO_2 separation performance with a selectivity up to 42; superior to that of many other ZIF-8 membranes.

Although polycrystalline metal-organic framework membranes have several advantages over other nanoporous membranes, they have not yielded the good CO_2 separation results which are essential for energy-efficient carbon capture. One of the most popular metal-organic frameworks, ZIF-8, has a crystallographically determined pore aperture of 0.34nm which is perfect for CO_2/N_2 and CO_2/CH_4 separation but its flexible lattice restricts the corresponding separation selectivities to below 5. A post-synthesis heat treatment, performable within a few seconds at 360C, markedly improved[202] the carbon-capture performance of ZIF-8 membranes. Lattice-stiffening was signalled by the appearance of a temperature-activated transport which was attributed to stronger interactions of the gas molecules with the pore aperture. The activation energy increased with molecular size: $CH_4 > CO_2 > H_2$. The CO_2/CH_4, CO_2/N_2 and H_2/CH_4 selectivities exceeded 30, 30 and 175, respectively, with complete blockage of C_3H_6 occurring. Spectroscopic and X-ray diffraction showed that the coordination environment and crystallinity were unaffected, while lattice distortion and strain were incorporated into the ZIF-8 lattice and increased its stiffness.

Using the robust metal-organic framework, ZIF-8, as the host and a flexible aliphatic alkylpolyamine, tetraethylenepentamine as the guest, an encapsulation compound, TEPA-ZIF-8, was prepared[203] via the insertion of tetraethylenepentamine into the pores of ZIF-8. Each cage in the ZIF-8 was occupied by some 1.44 tetraethylenepentamine molecules, and the introduced tetraethylenepentamine further adsorbed H_2O and CO_2 from the air so as to constitute a good proton conductor, TEPA-ZIF-8-H_2CO_3, with $\sigma = 2.08 \times 10^{-3}$S/cm at 293K and 99% relative humidity; with excellent durability of proton conduction. In the case of ZIF-8, the proton conductivity of TEPA-ZIF-8-H_2CO_3 increased by 3 orders of magnitude under the same conditions, and the activation energy decreased by 0.91eV.

The new concept of a core-shell type ionic liquid/metal organic framework composite was considered. A hydrophilic ionic liquid, 1-(2-hydroxyethyl)-3-methylimidazolium dicyanamide, [HEMIM][DCA], was deposited[204] onto a hydrophobic zeolitic imidazolate framework, ZIF-8. The composite exhibited an approximately 5.7 times higher CO_2 take-up and 45 times higher CO_2/CH_4 selectivity at 1mbar and 25C, as compared with that of the parent metal-organic framework. Ionic liquid molecules, deposited onto the external surface of the metal-organic framework, formed a core-shell type of material, in which the ionic liquid acted as a so-called smart gate for guest molecules.

Controlled encapsulation of atomically precise nanoclusters into metal-organic frameworks is an efficient method for creating new types of multifunctional crystalline porous material. So-called electrostatic attraction methods could be used to synthesize catalysts for the conversion of CO_2. The method was demonstrated[205] by preparing catalysts, including combinations of $[Au_{12}Ag_{32}(SR)_{30}]^{4-}$, $[Ag_{44}(SR)_{30}]^{4-}$ and $[Ag_{12}Cu_{28}(SR)_{30}]^{4-}$ nanoclusters with ZIF-8 and ZIF-67 frameworks. A $Au_{12}Ag_{32}(SR)_{30}$-ZIF-8 composite excelled in capturing CO_2 and converting phenylacetylene into phenylpropiolate under mild conditions (50C, ambient CO_2 pressure) with a turnover number as high as 18164. This far exceeded that of known catalysts. The catalyst was moreover very stable and could be re-used 5 times with no loss of catalytic activity.

A novel Lewis acid-base bifunctional Zn^{II}-based MOF-Zn-1 [$Zn_2L_2MA \cdot 2DMF$] (MA = melamine, H_2L = 2,5-thiophenedicarboxylic acid), with abundant micropores and free - NH_2 groups was easily assembled[206] by incorporating zinc^{II} ion with nitrogen-rich melamine and 2,5-thiophenedicarboxylic acid ligands. The constructed MOF-Zn-1 had a high affinity for CO_2 molecules due to the Lewis-base property together with abundant micropores. The Zn active sites could be used for epoxide activation. The acid-base synergistic effects facilitated CO_2 conversion into cyclic carbonates under ambient temperature using the porous MOF-Zn-1 as a heterogeneous catalyst. Moreover, the MOF-Zn-1 exhibited stability and versatility, and it was easy to recycle with no obvious decrease in catalytic activity.

Mixed-ligand 3D/2D zinc metal–organic frameworks, $\{[Zn(bdc)(L1)]_xG\}_n$ (ZnMOF-1) and $\{[Zn(ipa)(L2)]\}_n$ (ZnMOF-2; in which H_2BDC = benzene-1,4-dicarboxylic acid, L1 = 4-pyridyl carboxaldehyde isonicotinoylhydrazone, H_2IPA = isophthalic acid, L2 = 3-pyridyl carboxaldehyde nicotinoyl hydrazone and G = lattice guests) were synthesized[207] using versatile synthetic routes that included a green mechanochemical (grinding) reaction. Chemical and thermal stability, phase purity, and characterization of the ZnMOFs synthesized using different approaches were established by using various analytical methods. Both ZnMOFs can be used as a highly active, solvent-free, binary catalyst for CO_2 cyclo-addition with epoxides under ambient reaction conditions of 1atm pressure and room temperature/40C, in the presence of the co-catalyst, nBu_4NBr. The yield, recyclability, and stability of ZnMOF-1 as a potential catalyst towards epoxide to cyclic carbonate conversion are excellent under ambient conditions. From literature and experimental inferences, a rationalized mechanism mediated by the Zn center of ZnMOFs for the CO_2-epoxide cyclo-addition reaction has been proposed. Very few MOF-based catalysts seem to have been reported for the conversion of CO_2 to useful products under similarly mild conditions. In the present investigation, that is, catalyst preparation by green mechanochemical synthesis and catalysis under ambient, solvent-free conditions were performed with minimum energy use.

Isostructural nano Zn^{II}-based metal–organic frameworks, Zn-TMU, have been synthesized by ultrasound processing and solvent-assisted linker exchange. Zinc^{II} metal-organic frameworks were investigated as CO_2 capture compounds. Compared with the ultrasound process, the as-prepared daughter frameworks showed an enhanced CO_2 sorption capacity, attributed to the existence of structural defects during solvent-assisted linker exchange. This demonstrated[208] that gas storage depended mainly upon the quality and defects in the structure, which depended in turn upon the preparation conditions.

A dynamic porous coordination polymer material with local flexibility, in which the propeller-like ligands rotate to permit CO_2 trapping, was described[209]. Owing to its high affinity for CO_2 and the confinement effect, the porous coordination polymer exhibits high catalytic activity, rapid transformation dynamics, and high size-selectivity for different substrates. Together with an excellent stability, with turnover numbers of up to 39000 per $Zn_{1.5}$ cluster of catalyst after 10 cycles for CO_2 cyclo-addition to form value-added cyclic carbonates, these results demonstrate that such a distinctive structure is responsible for visual CO_2 capture and size-selective conversion.

A trigonal nanosized carboxylate ligand, 4, 4′,4″-(triazine-2,4,6-triyl-tris(benzene-4,1-diyl))tribenzoic acid (H_3TAPB), has been used[210] for the construction of a highly porous metal−organic framework. A solvothermal reaction of H_3TAPB and a zinc salt in a mixed solvent of DEF/dioxane/H_2O produced the targeted porous MOF $[Zn_3(TAPB)_2(H_2O)2](DEF)_4(H_2O)_3$ (1, DEF = N,N-diethylformamide). This MOF contains a linear Zn_3 secondary building unit with two terminal water molecules occupying the axial positions, which is connected by the TAPB3− ligands to afford a 3-fold interpenetrated framework with a large calculated void volume (69.7%). The activated MOF shows exceptional porosities and high CO_2 take-up capacities as well as good sorption selectivity of CO_2/CH_4 at around room temperature.

An iso-reticular metal-organic framework-3(IRMOF-3) exhibiting microporosity was synthesized[211] from organic linker 2-amino terephthalic acid and the salt zinc nitrate hexahydrate ($Zn-(NO_3)_2 \bullet 6H_2O$) via catalyst-free, solution-based direct mixing. Three samples of IRMOF-3 were post-synthetically functionalized with aminomethyl propanol (AMP) at three different concentrations, 25, 50 and 75wt%, to produce AMP-IRMOF-3. The structural and texture properties of AMP-IRMOF-3 were studied by FESEM, FTIR, pore size distribution, powder XRD, TGA/DTG, N_2 adsorption-desorption isotherms, and their selective CO_2/CH_4 adsorption behavior at various temperatures (298.15, 323.15 and 348.15K). The results showed an enhancement in CO_2 adsorption capacity from 1.39 to 3.90mmol/g due to successful incorporation of CO_2 philic functionalities. Furthermore, the adsorption isotherms were studied by applying Langmuir, Freundlich, Langmuir-Freundlich, and Toth isotherm models. The isotherm study revealed an unfavourable adsorption behavior, with an increase in AMP loadings for an homogeneous system.

A series of new zinc^II-thiophene-2,5-dicarboxylate (tdc) MOFs based upon novel dodecanuclear wheel-shaped building blocks has been synthesized[212]. Single-crystal X-ray diffraction analyses reveal 3-dimensional porous frameworks with a complex composition $[Zn_{12}(tdc)_6(glycolate)6(dabco)_3]$ where glycolate is a de-protonated polyatomic alcohol (ethylene glycol, EgO_2, 1; 1,2-propanediol, PrO_2, 2; 1,2-butanediol, BuO_2, 3; 1,2-pentanediol, PeO_2, 4; glycerol, GlO_2, 5) and dabco is 1,4-diazo[2.2.2.]bicyclooctane. All of the compounds were isostructural apart from pendant groups of the diols decorating the surface of channels. The adsorption of small gases (N_2,

CO_2, CH_4, C_2H_2, C_2H_4, C_2H_6) and larger hydrocarbons (benzene, cyclohexane) in both liquid and vapor phases was thoroughly investigated for all compounds. The zero-coverage adsorption enthalpies, Henry constants, and selectivity factors required by various models were calculated. The versatile adsorption functionality of the title series results from the variable nature of the diol and could be tailored for a specific adsorbate system. For example, 1 shows excellent selectivity of benzene over cyclohexane (20:1 for vapors, 92:1 for liquid phase), while 4 demonstrates an unprecedented adsorption preference of cyclohexane over benzene (selectivity up to 5:1). The compound 5 demonstrates great adsorption selectivity for CO_2/N_2 (up to 75.1), CO_2/CH_4 (up to 7.7), C_2H_2/CH_4 (up to 14.2), and C_2H_4/CH_4 (up to 9.4). Also, due to the polar nature of the pores, 5 features size-selective sorption of alkaline metal cations in the order, $Li^+ > Na^+ > K^+ > Cs^+$, as well as a notable luminescent response for cesium[I] ions and urea.

Gas adsorption and diffusion were investigated in an emerging pillared-bilayer metal-organic framework, Zn-AIP-AZPY (aip, 5-aminoisophthalic acid; azpy, 4,4′-azobipyridine). This work demonstrated that the crystal structure of this flexible MOF could be controlled by two different activation methods: thermal activation and chemical activation. The pore limiting diameter of the thermally activated compound was about 3.7Å, whereas that of the fully activated one formed via methanol extraction was 2.92Å. Although the aperture size of the fully activated structure was smaller[213] than the kinetic diameter of CO_2 and CH_4, Zn-AIP-AZPY still presented a fair degree of adsorption to these two gases. *In situ* XRD experiments under various gases at various pressures also suggested a minor breathing effect when Zn-AIP-AZPY was exposed to CO_2 but no breathing effect under CH_4. These results imply the possibility of ligand rotation occurring during the adsorption of CO_2 and CH_4 in Zn-AIP-AZPY. The Zn-AIP-AZPY membranes were further prepared using the seeded growth method, and these membranes showed an exceptionally high H_2 permeability (over 105 barrer) with a good H_2/CO_2 selectivity (ideal selectivity beyond 8). A reverse CO_2/CH_4 selectivity, i.e., CH_4 permeates faster than CO_2, was surprisingly found. Monte Carlo simulations were conducted for probing the adsorption properties of CO_2 and CH_4, as well as their adsorption energy landscape in Zn-AIP-AZPY. It was found that this MOF energetically favoured the adsorption of CO_2 over CH_4 by nearly 10kJ/mol.

Ligand exchange reactions at the Kuratowski-type secondary building unit in MFU-4l(arge) metal-organic frameworks result[214] in organometallic porous compounds with metal-carbon bonds with the general formula $[Zn_5L_xCl_{4-x}(BTDD)_3]$ ($4 \geq x > 3$; L = methanido, ethanido, n-butanido, tert-butanido, 3,3-dimethyl-1-butyn-1-ido; H_2-BTDD = bis (1H-1,2,3-triazolo [4,5-b] [4′,5′-i]) dibenzo [1,4] dioxin) and $[Zn_{1.5}Co_{3.5}Me_{3.1}Cl_{0.9}(BTDD)_3]$. The hydrolytic stability of every compound was examined, and a conversion of the methanide to hydroxide ligands was observed in the cobalt-containing compound. DRIFTS measurements of the resulting framework with the composition $[Zn_{1.4}Co_{3.6}(OH)_{3.1}Cl_{0.9}(BTDD)_3]$ revealed a mechanism of carbon dioxide binding similar to that of carbonic anhydrase.

A double interlacing metal-organic framework, with the formula, $[(Zn_4O)_2(PDDA)_6(H_2O)_2]\cdot10DMF$, with pcu-topology has been assembled[215] from Zn_4O inorganic clusters and a V-shaped ligand 4,4'-(pyridine-2,6-diyl) dibenzoic acid (denoted as H_2PDDA) under solvothermal conditions. Benefiting from its internal porosity, with available Lewis basic sites and open metal centers, the compound exhibits an excellent performance with regard to CO_2 transformation with epoxides and also selective luminescence sensing for nitrofuran antibiotics. It can therefore be used as an efficient bifunctional platform as both a catalyst for CO_2 cyclo-addition reactions and as a sensor for antibiotic detection.

Two new adenine-based Zn^{II}/Cd^{II} metal-organic frameworks have been synthesized: $[Zn_2(H_2O)(stdb)_2(5H\text{-}Ade)(9H\text{-}Ade)_2]_n$ (PNU-21) and $[Cd_2(Hstdb)(stdb)(8H\text{-}Ade)(Ade)]n$ (PNU-22), containing auxiliary dicarboxylate ligand (stdb = 4,4'-stilbenedicarboxylate). Both MOFs were structurally robust and possessed unsaturated Lewis acidic metal centers [Zn^{II} and Cd^{II}] and free basic N atoms in adenine molecules[216]. They were used as heterogeneous catalysts for the fixation of CO_2 into 5-membered cyclic carbonates. Significant conversion of epichlorohydrin was attained at a low CO_2 pressure (0.4MPa) and moderate catalyst (0.6mol%)/cocatalyst (0.3mol%) amounts, with over 99% selectivity towards the epichlorohydrin carbonate. They showed comparable or even higher catalytic activity than other previously reported metal-organic frameworks. Because of a high thermal stability and robust architecture of PNU-21/PNU-22, both catalysts could be re-used with simple separation for up to five successive cycles without any considerable loss of their catalytic activity. Densely populated acidic and basic sites in both Zn^{II}/Cd^{II} metal-organic frameworks facilitated the conversion of epichlorohydrin to epichlorohydrin carbonate in high yields.

An adenine-dependent CO_2 photoreduction in green biomimetic metal–organic frameworks was demonstrated. Photocatalytic results indicated[217] that AD-MOF-2 exhibited a very high HCOOH production rate of 443.2μmol/gh in pure aqueous solution, and was more than 2 times higher than that of AD-MOF-1 (179.0μmol/gh) in acetonitrile solution. Experimental and theoretical evidence revealed that the CO_2 photoreduction reaction took place mainly at the aromatic nitrogen atom of adenine molecules, via an unique o-amino-assisted activation, rather than at the metal center.

Two metal–organic frameworks, $[Zn_2(atrz)_2(bpdc)]_{0.5}$ H_2O and $[Zn_2(mtrz)_2(azdc)]\cdot DMA\cdot CH_3OH\cdot H_2O$, with 3-substituted 1,2,4-triazole (Hatrz = 3-amino-1,2,4-triazole, Hmtrz = 3-methyl-1,2,4-triazole) have been constructed[218]. Both metal-organic frameworks were 3-dimensional self-penetrating frameworks featuring Zn–triazolate layers pillared by dicarboxylate ligands (H_2bdc = 1,4-benzenedicarboxylate, H_2azdc = azobenzene-4,4′-dicarboxylate). In particular, $[Zn_2(mtrz)_2(azdc)]\cdot DMA\cdot CH_3OH\cdot H_2O$ exhibited a rare rob network with the dimeric Zn–triazolate as the 6-connected nodes, which resulted from the synergistic effects of both 3-substituted group of triazolate and size-alterable dicarboxylate pillars upon the final frameworks. Detailed structural analysis of $[Zn_2(atrz)_2(bpdc)]_{0.5}H_2O$ and $[Zn_2(mtrz)_2(azdc)]\cdot DMA\cdot CH_3OH\cdot H_2O$ and a comparison with commonly observed pcu nets in pillared-layer structure was made. Pore-size tuning and functionality were related to the various molecular lengths of the pillars. Thus $[Zn_2(mtrz)_2(azdc)]\cdot DMA\cdot CH_3OH\cdot H_2O$, with long pillars, had a suitable pore size and hence exhibited selective CO_2 take-up and efficient adsorption in aqueous solutions, based upon the size-exclusion effect.

Three pillar-layered metal-organic frameworks based upon $M(HBTC)(4,4'-bipy)\cdot 3DMF$ (M = Ni, Co, and Zn; HBTC = 1,3,5-benzenetricarboxylic acid, 4,4'-bipy = 4,4'-bipyridine) were synthesized[219] using a solvothermal method. $Zn(HBTC)(4,4'-bipy)\cdot 3DMF$ was synthesized for the first time using both a solvothermal and microwave method, and subsequently characterized by various physicochemical methods. The structure of $M(HBTC)(4,4'-bipy)\cdot 3DMF$ consisted of honeycomb grid layers of M^{2+} ions and BTC units, which were further linked by the 4,4'-bipy pillars to form a highly porous 3-dimensional framework. All of the metal-organic frameworks displayed excellent synergistic catalytic properties with alkyl ammonium halides involved in the solvent-less fixation of CO_2 with epoxides to produce cyclic carbonates. The catalytic activities of these metal-organic frameworks followed the trend, Zn > Co > Ni, which was explained by the acid-base bifunctional properties. The microwave-synthesized $Zn(HBTC)(4,4'-bipy)\cdot 3DMF$ material exhibited physical, chemical, and catalytic properties that were similar to those of the catalyst obtained using conventional solvothermal synthesis.

Two metal-organic framework isomers with the chemical formula $Zn_2(X)_2(DABCO)$ [X = terephthalic acid (BDC), dimethyl terephthalic acid (DM), 2-aminoterephthalic acid (NH_2), 2,3,5,6-tetramethyl terephthalic acid, and anthracene dicarboxylic acid (ADC); DABCO = 1,4-diazabicyclo[2.2.2]octane] have been synthesized[220] via a fast room-temperature synthesis route. The synthesis solvent was found to play a vital role in directing the formation of a Kagome lattice (ZnBD) versus tetragonal topology (DMOF-1). When N,N-dimethylformamide or dimethyl sulphoxide was used as the synthesis solvent, the reaction resulted in the formation of ZnBD, whereas methanol, ethanol, acetone, N,N-diethylformamide, and acetonitrile each produced DMOF-1. Water adsorption isotherms of ZnBD and DMOF-1 were collected, and the materials were found to have similar adsorption characteristics and stabilities. Both metal-organic frameworks degraded upon exposure to water at a relative pressure (P/P_o) of 0.5 at 25C, but both are hydrophobic below a P/Po of 0.4, displaying very little water adsorption. Additionally, CO_2 adsorption isotherms of ZnBD were collected and compared to those previously reported for DMOF-1. ZnBD adsorbs less CO_2 at low pressure as compared with DMOF-1 but reaches a similar capacity at 20bar. This adsorption behavior can be explained by the structural features of the materials, where ZnBD possesses large hexagonal pores (15Å) as compared to the smaller pore opening (7.5Å) in DMOF-1. The heat of adsorption of CO_2 on ZnBD was calculated to be 22kJ/mol at zero coverage. Attempts to functionalize the Kagome lattice proved to be unsuccessful but instead resulted in a new method for producing functionalized DMOF-1 at room temperature. This was hypothesized to be a result of the steric effects imposed by the functional groups that prevent the formation of the Kagome lattice.

A 2-dimensional Zn-MOF, $\{Na[Zn_{1.5}(\mu_4\text{-}O)(L)]\}_n$, was synthesized[221] under hydrothermal conditions and characterized by single-crystal X-ray diffraction. Four zinc atoms were bridged through μ_4-O to form $[Zn_4O]$ clusters, which were further linked to form a 2-dimensional layer network by sharing zinc atoms at vertices. It exhibited a high thermal stability up to 280C and remained stable in common solvents and aqueous solutions with pH values ranging from 1 to 13. Catalytic studies revealed that the material exhibited excellent catalytic activity for cyclo-addition of CO_2 with epoxides into cyclic carbonates under mild conditions. It exhibited good generality in CO_2 coupling reactions with many epoxides and could be re-used at least 5 times without significant reduction in catalytic ability.

Four types of novel jungle-gym-type porous coordination polymers, $[Zn_2(bdc)_2(dabco)]n$ (Zn, bdc = 1,4-benzenedicarboxylate, dabco = 1,4-diazabicyclo[2.2.2]octane), $[Zn_2(bdc\text{-}NO_2)_2(dabco)]_n$ (Zn-NO_2, bdc-NO_2 = nitro-terephthalate), $[Cu_2(bdc)_2(dabco)]_n$ (Cu) and $[Cu_2(bdc\text{-}NO_2)_2(dabco)]_n$ (Cu-NO_2), were synthesised[222] and characterised using gas adsorption measurements to estimate the affinity between CO_2 and nitro moieties. The use of N_2 adsorption measurements at 77K revealed type-I isotherms, indicating permanent porosities, where the Brunauer–Emmett–Teller surface areas were 1417m^2/g (Zn), 1076m^2/g (Zn-NO_2), 1688m^2/g (Cu), and 759m^2/g (Cu-NO_2). The adsorption capacity of CO_2 at 195K parallels the amount of surface area, whereas the isosteric heat of CO_2 adsorption in both Zn-NO_2 and Cu-NO_2 are higher than for the original compounds. The enthalpies at zero coverage of CO_2 adsorption are 19kJ/mol (Zn),

27kJ/mol (Zn-NO$_2$), 17kJ/mol (Cu), and 27kJ/mol (Cu-NO$_2$), which can be attributed to the quadruple interaction between CO$_2$ and nitro moieties.

A Zn^{II}-based metal–organic framework, [Zn$_3$(TCA)$_2$(DPE)]·DMF·6H$_2$O, BUT-161, was solvothermally synthesized[223] by the reaction of a tritopic carboxylic acid 4,4′,4″-nitrilotribenzoic acid (H$_3$TCA), an ancillary ditopic pyridine ligand 1,2-di(4-pyridyl)ethylene (DPE) and Zn(NO$_3$)$_2$•6H$_2$O. The framework of BUT-161 had a pillar-layered structure, where 2-dimensional layers constructed from trimeric Zn$_3$(–CO$_2$)$_6$ clusters and TCA3− ligands were a topologically (3,6)-connected net, and the layers were further pillared by the DPE linkers. The resulting 3-dimensional framework contained 1-dimensional channels along the b-axis. Powder X-ray diffraction patterns and gas adsorption studies revealed that the activated phase of BUT-161, as BUT-161a, had a permanent porosity with a Brunauer-Emmett-Teller specific surface area of 308m^2/g, although the framework structure shrank (structure transformation) following guest removal. It was found that BUT-161a could selectively adsorb CO$_2$ over N$_2$ or CH$_4$.

A new metal-organic framework {[Zn$_2$(Htzba)$_2$(dmtrz)]·(CH$_3$)$_2$NH}$_n$ that had already been solvothermally synthesized exhibited[224] a high CO$_2$ take-up capacity based upon experimental and simulated results. The CO$_2$/CH$_4$ and CO$_2$/N$_2$ selectivities were predicted to be 41.19 and 46.21 at 298K and 100kPa, respectively. The superior performance suggested that this material is a promising candidate for CO$_2$ separation.

Two amino functionalized metal-organic frameworks, {[Zn(Py$_2$TTz)(2-NH$_2$-BDC)]·(DMF)}$_n$ and {[Cd(Py$_2$TTz)(2-NH$_2$-BDC)]·(DMF)·0.5(H$_2$O)}$_n$ where Py$_2$TTz = 2,5-bis(4-pyridyl)thiazolo[5,4-d]thiazole, 2-NH$_2$-BDC = 2-amino-1,4-benzenedicarboxylate, and DMF = N,N-dimethylformamide), were synthesized[225] and characterized using the primary ligand 2-amino-1,4-benzenedicarboxylic acid (2-NH$_2$-H$_2$BDC) and the auxiliary ligand 2,5-bis(4-pyridyl)thiazolo[5,4-d]thiazole (Py2TTz). They possessed similar 2-fold interpenetrated 3-dimensional bipillar-layer framework structures composed of typical binuclear metal nodes, 2-NH$_2$-BDC two-dimensional layers and Py$_2$TTz bipillars. It was notable that thiazole nitrogen atoms and pendant -NH$_2$ groups were present in channels in the two frameworks. Given their good chemical stabilities, high thermal stabilities, and exposed nitrogen sites, gas adsorption and catalytic experiments were performed on the two metal-organic frameworks. The results demonstrated that MOF 2 could selectively absorb carbon dioxide gas and that the two metal-organic frameworks could be used as recyclable heterogeneous catalysts for Knoevenagel condensation reactions under solvent-free conditions.

A double-walled metal-organic framework, [Zn$_5$(μ_3-OH)$_2$(DBTA)$_2$(H$_2$O)$_4$]solvents [1, H4DBTA = 2,2′-dihydroxy-1,1′-binaphthyl-3,3′,6,6′-tetrakis(4-benzoic acid)], was prepared[226], based upon a rare Zn_5 cluster. The resultant sample has an enrichment of the porous structure, with a high Brunauer–Emmett–Teller and Langmuir surface area. Because of the multi-aperture structure and plentiful catalytic sites of Brønsted (-OH) and Lewis acidic (Zn^{II}), MOF 1 is a unique catalyst for the synthesis of cyclic carbonates from CO_2 and epoxides with preferred repeatability.

Three components of pillared metal-organic frameworks, metal ion, carboxylic acid ligand and N-chelating ligand, were controlled for CO_2 cyclo-addition catalysts used to synthesize organic cyclic carbonates. Among the divalent metals, Zn^{2+} showed the best catalytic activity, and in DABCO (1,4-diazabicyclo[2.2.2]octane)-based metal-organic frameworks, hydroxy-functionalized DMOF-OH was the most efficient MOF for CO_2 cyclo-addition. For the BPY (4,4'-bipyridyl)-type metal-organic frameworks, all 5 prepared BMOFs (BPY MOFs) showed[227] a similar and good conversion of CO_2 cyclo-addition. This pillared metal-organic framework could be recycled up to 3 times without activity or crystallinity loss.

Three new 3-dimensional porous Zn^{II}-based metal-organic frameworks, $[Zn_4O(L)_2(NMP)_2(H_2O)]\cdot 2NMP\cdot 2H_2O$, $[Zn(HL)$ $(bpe)_{0.5}]\cdot DMF\cdot H_2O$ and $[Zn(HL)$ $(bipy)_{0.5}]\cdot DMF\cdot H_2O$ [bpe = 1,2-di(pyridin-4-yl)ethene, and bipy = 4,4'-bipyridine], were synthesized[228] via 5'-carboxyl-(1,1'-3',1''-terphenyl)-4,4''-dicarboxylic acid (H3L). Single-crystal X-ray diffraction shows that $[Zn_4O(L)_2(NMP)_2(H_2O)]\cdot 2NMP\cdot 2H_2O$ is a 2-fold interpenetrated 3-dimensional framework incorporating $[Zn_4O(COO)_6]$ secondary building units, while $[Zn(HL)(bpe)_{0.5}]\cdot DMF\cdot H_2O$ and $[Zn(HL)(bipy)_{0.5}]\cdot DMF\cdot H_2O$ [bpe = 1,2-di(pyridin-4-yl)ethene, and bipy = 4,4'-bipyridine] are 3-dimensional isostructural networks with different N-donor ancillary ligands, wherein partially de-protonated HL2-ligands are included.

A new anionic metal-organic framework, $\{[H_3O]_2[Zn_4(\mu_4\text{-}O)(NSBPDC)_4]\cdot 16(H_2O)]\}_n$, was synthesized[229] under solvothermal conditions using a pre-designed multifunctional dicarboxylic acid ($H_2NSBPDC$ = 6-nitro-2,2'-sulfone-4,4'-dicarboxylic acid). The complex contained 4-connected tetrahedral $[Zn_4(\mu_4\text{-}O)(CO_2)_8]$ clusters, which were further linked by the bridging ligands; generating a threefold interpenetrated diamond-like network with ultra-microporous channels. Gas adsorption studies revealed that the compound possessed a good adsorption selectivity for CO_2 over CH_4 and N_2. In addition, it could serve as a host for the incorporation of lanthanide cations via targeted ion-exchange. Its structure could be reversibly dehydrated and rehydrated.

Two Zn-based metal–organic frameworks of pyridinemethanol–carboxylate conjugated ligands, namely, $[Zn(L1)]_n xSol$ (1, 3D) and $[Zn(L2)_2]_n$ (2, 1D) (H_2L1 = 4-(6-(hydroxymethyl)pyridin-3-yl)benzoic acid; H_2L_2 = 3-(6-(hydroxymethyl)pyridin-3-

yl)benzoic acid) have been synthesized[230] and structurally characterized. The dimensionalities of 1 and 2 were defined by the de-protonation states of the ligands. In particular, the 3-dimensional MOF 1 featured a rod-shaped Zn–O/COO chain as the secondary building unit which effectively hindered network interpenetration, whereas the 1-dimensional chain of 2 mimicked an edge-sharing octahedron with each Zn center serving as the vertex. MOF 1 reversibly took up CO_2 and had a Brunauer–Emmett–Teller surface area of $345m^2/g$, and remained crystalline upon activation; an indication of permanent porosity.

The complex, $[Zn_2(tdc)_2dabco]$ (H_2tdc = thiophene-2,5-dicarboxylic acid; dabco = 1,4-diazabicyclooctane), exhibits[231] a remarkable increase in carbon dioxide take-up and CO_2/N_2 selectivity as compared to the non-thiophene analogue $[Zn_2(bdc)_2dabco]$ (H_2bdc = benzene-1,4-dicarboxylic acid; terephthalic acid). The CO_2 adsorption at 1bar for $[Zn_2(tdc)_2dabco]$ is $67.4cm^3/g$ (13.2wt%) at 298K and $153cm^3/g$ (30.0wt%) at 273K. For $[Zn_2(bdc)_2dabco]$, the equivalent values are $46cm^3/g$ (9.0wt%) and $122cm^3/g$ (23.9wt%), respectively. The isosteric heat of adsorption for CO_2 in $[Zn_2(tdc)_2dabco]$ at zero coverage is low (23.65kJ/mol), ensuring easy regeneration of the porous material. Enhancement by the thiophene group of the separation of CO_2/N_2 gas mixtures has been confirmed by both ideal adsorbate solution theory calculations and dynamic breakthrough experiments. The preferred binding sites of adsorbed CO_2 in $[Zn_2(tdc)_2dabco]$ have been unambiguously determined by *in situ* single-crystal diffraction studies of CO_2-loaded $[Zn_2(tdc)_2dabco]$, coupled with quantum-chemical calculations. These studies revealed the role of the thiophene moieties in the specific CO_2 binding via an induced dipole interaction between CO_2 and the sulfur center, confirming that an enhanced CO_2 capacity in $[Zn_2(tdc)_2dabco]$ is achieved without the presence of open metal sites. The experimental data and theoretical insight suggest a viable strategy for the improvement of the adsorption properties of already-known materials through the incorporation of sulfur-based heterocycles within their porous structures.

By appealing to the pillared-layer strategy, a new Zn^{II}-based microporous polycatenated metal-organic framework $\{[Zn_2(bpydb)_2(EtOH)(4\text{-}bpdb)](DMF)_4(EtOH)_2\}_n$ (1, bpydb = 4,4'-(4,4'-bipyridine-2,6-diyl)dibenzoate, 4-bpdb = 1,4-bis(4-pyridyl)-2,3-diaza-1,3-butadiene, DMF = N,N-dimethylformamide) with Lewis basic sites on the pore surface was prepared[232] and generated by the solvothermal reaction of mixed ligands (bpydbH$_2$ and 4-bpdb) and $Zn(NO_3)_2 \bullet 6H_2O$. The structure of 1 exhibited a 3-dimensional pillared-bilayer network, generated by polycatenation of the 2-dimensional bilayers with an accessible solvent void of 43%. Its porosity has been verified by N_2 sorption at 77K, which reveals a Brunauer–Emmett–Teller surface area of $545m^2/g$ and a pore volume of $0.324cm^3/g$. Moreover, compound 1 shows an excellent adsorption selectivity for CO_2

over CH_4 at around room temperature, which has been correlated with the specific interaction of CO_2 with the functional azine groups that decorated the pore channels.

Reactions of Zn^{II} ions with various coordination modes and bis(4-carboxy-benzyl)amine (H_2bcba) with a secondary aliphatic amine bridge at different temperatures yielded[233] a pair of isomeric frameworks with very similar framework structures but distinctly different local structures, namely α-[Zn(bcba)] (MCF-51 or α-Zn), with all amino groups being coordinated and β-[Zn(bcba)] (MCF-52 or β-Zn) with uncoordinated amino groups. Single-component gas adsorption and CO_2/CH_4 mixture breakthrough experiments under ambient conditions showed that the CO_2 adsorption ability and CH_4 purification performances of β-Zn, in terms of CO_2 adsorption enthalpy (42kJ/mol), CO_2/CH_4 selectivity (32) and CH_4 purity/productivity (1.49mmol/g at 99.999%+), were much higher than those of α-Zn (CO_2 adsorption enthalpy of 17kJ/mol, CO_2/CH_4 selectivity of 2.9 and CH_4 purity/productivity of 0.088mmol/g at 99%+).

A flexible dipyridyl ligand 1,2-bis(4-pyridylmethylamino)ethane (L) was synthesized[234] and its various coordination modes with Zn^{II} and Cd^{II} in the presence of divers dicarboxylate yielded five novel metal–organic frameworks, $\{[Zn(L)(BDC)]\cdot_{6.5}H_2O\}_n$ (1), $\{[Zn(L)(NH_2\text{-}BDC)]\cdot_{6.5}H_2O\}_n$ (2), $\{[Cd(L)(NH_2\text{-}BDC)]\ 7H_2O\}_n$ (3), $\{[Cd(L)(OH\text{-}BDC)]\ H_2O\}_n$ (4), and $\{[Cd(L)(diOH\text{-}BDC)]\ 2H_2O\}_n$ (5) (where, L= 1,2-bis(4-pyridylmethylamino)ethane and BDC = benzene-1,4-dicatboxylic acid). Crystal-structure analysis revealed that compounds 1 and 2 were 3-dimensional porous networks with open channels while compounds 3 to 5 were 2-dimensional metal-organic frameworks. Complex-3 contained hydrogen-bonded water molecules between the 2-dimensional layers while 4 and 5 were high-density compounds. The ligand L adopted various conformations which governed the final topology of the complexes. Complexes 1 and 2 had the rare topology, gra, while 3 to 5 had four connected uninodal nets with sql topology. The solvent-accessible void volumes of 1 to 3 were 51.2, 52.3 and 27%, respectively. Gas-adsorption studies revealed that 1 to 3 exhibited permanent porosities having type-I isotherms, and selectively adsorbed CO_2 gas over nitrogen and methane at low temperatures, with final take-ups of 24.18, 22.35 and 14.64cc/g at STP, respectively. Vapour sorption studies revealed that all of the complexes selectively adsorbed water vapour over methanol or ethanol vapor, with the final amounts reaching 182.44, 219.29, 189.21, 81.79 and 66.48cc/g at room temperature.

A novel bottom-up strategy was reported[235] for the direct growth of a highly oriented Zn_2(bIm)$_4$ (bIm = benzimidazole) ZIF nanosheet tubular membrane, based upon graphene oxide (GO) guided self-conversion of ZnO nanoparticles. Using this approach, a thin layer of ZnO nanoparticles confined between a substrate and a GO ultra-thin layer self-converts into a highly oriented Zn_2(bIm)4 nanosheet membrane. The resulting membrane with a thickness of around 200nm exhibits an excellent H_2/CO_2 gas separation performance with a H_2 performance of 1.4 x 10^{-7}mol/m^{2}sPa and an ideal separation selectivity of about 106. The method can be easily scaled up and extended to the synthesis of other types of Zn-based metal-organic framework nanosheet membranes.

Metal-organic frameworks, M(BPZNH$_2$) (M = Zn, Ni, Cu), have been prepared[236] by reacting the corresponding metal acetates M(OAc)$_2$nH$_2$O and the organic linker 3-amino-4,4′-bipyrazole (H$_2$BPZNH$_2$) under solvothermal conditions. The H$_2$BPZNH$_2$ was obtained straightforwardly from the reduction of the related nitro-compound, using hydrazine as a reducing agent. The Zn^{II} polymer was characterized by a 3-dimensional porous network featuring tetrahedral metallic nodes and bridging BPZNH$_2$ 2-anions defining the vertices and edges of square channels. The isostructural Ni^{II} and Cu^{II} metal-organic frameworks had square-planar metallic nodes and bridging BPZNH$_2$ 2-spacers at the vertices and edges of the rhombic channels of a 3-dimensional porous framework. They are micro-mesoporous materials with Brunauer–Emmett–Teller specific surface areas of 100 to 400m^2/g. The Zn(BPZNH$_2$) had the highest area (395m^2/g) and a mainly microporous texture (micropore area = 69% of the accessible SSA), was used as a CO_2 capture material: at 298K and 1bar; the total gas take-up equalling 3.07mmol/g (13.5wt%CO_2). Its affinity for CO_2 (isosteric heat of adsorption Qst = 35.6kJ/mol; CO_2/N_2 Henry selectivity = 17; CO_2/N_2 IAST selectivity = 14) is higher than that of its nitro-functionalized analogue and comparable to that of known amino-decorated metal-organic frameworks. The Zn(BPZNH$_2$) was also tested as an heterogeneous catalyst in the reaction of CO_2 with activated epoxides having a -CH$_2$X pendant arm (X = Cl: epichlorohydrin; X = Br: epibromohydrin) to give the corresponding cyclic carbonates at 393K and 1bar under green (solvent- and co-catalyst-free) conditions. Good conversion (47%) and a TOF of 3.9mmol(carbonate)/(mmolZn)h were recorded for epibromohydrin.

The ionic Zn^{II}-porphyrin complex, [Zn^{II}NMTPyP]$^{4+}$[I^{-}]$_4$ where Zn^{II}NMTPyP is 5,10,15,20-tetrakis(1-methylpyridinium-4′-yl) zinc^{II} porphyrin, was immobilized[237] in the porous metal-organic framework, PCN-224. Calculations confirmed incorporation of the complex into the 1-dimensional channels of PCN-224 due to a suitable matching of the pore size to the size of the complex, and to stabilization by π-π interactions. The hybrid exhibited a marked affinity for CO_2, with a high (40.5kJ/mol) heat of adsorption as compared with that (36.8kJ/mol) of the pristine PCN-224. The hybrid acted as an efficient environmentally-friendly co-catalyst and solvent-free heterogeneous catalyst for the fixation of CO_2 and the generation of cyclic carbonates. Porphyrin-based [Zn$_3$(C$_{40}$H$_{24}$N$_8$)(C$_{20}$H$_8$N$_2$O$_4$)$_2$(DEF)$_2$](DEF)$_3$, where DEF is N,N-diethylformamide, was synthesized[238] from 5,10,15,20-tetrakis(4-pyridyl)porphyrin, 1,2-diamino-3,6-bis(4-carboxyphenyl)benzene and a Zn^{2+} salt at 100C under solvothermal conditions. The structure was 3-dimensional, with three-fold interpenetration. Free -NH$_2$ groups in the ligand anchored silver nanoparticles via simple solution, and silver nanoparticles with a size of 3.83nm were uniformly distributed throughout the metal–organic framework. Porphyrin-based metal-organic frameworks were prepared using zinc, aluminium or cobalt[239]. The aluminium-based material produced high (4.3%) CO$_2$ photoreduction conversion. Further study indicated that the maximum (10.63%) photoreduction conversion of CO$_2$ under optimum conditions would occur with 297.24mg of catalyst, a total feed pressure of 1.4atm and with methanol as a sacrificial component. Ultra-thin 2-dimensional zinc porphyrin-based metal–organic framework nanosheets have been used[240] for the photoreduction of CO$_2$ to CO. These nanosheets exhibited better charge-transport capabilities and longer photogenerated electron-hole pair lifetimes than did the bulk material.

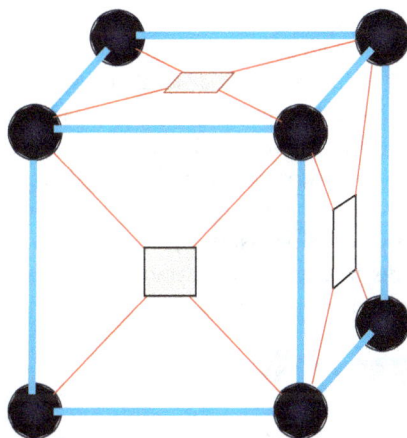

Figure 13 The ftw topology

Zirconium-

A novel series of two zirconium- and one indium-based metal-organic frameworks, MOF-892, MOF-893, and MOF-894, constructed from the hexatopic linker, 1′,2′,3′,4′,5′,6′-hexakis(4-carboxyphenyl)benzene, were synthesized[241]. The MOF-892 and MOF-893 were new exemplars of materials having topologies which were previously unknown in the family of zirconium metal-organic frameworks. The MOF-892, MOF-893 and MOF-894 exhibited efficient heterogeneous catalytic activity for the cyclo-addition of CO$_2$; resulting in cyclic organic carbonate formation with high conversion, selectivity and yield under mild conditions (1atm CO_2, 80C, solvent-free). Because of the structural features provided by their building units, MOF-892 and MOF-893 were replete with accessible Lewis and Brønsted acid sites located at the metal clusters and the non-coordinating carboxylic groups of the linkers, respectively. This is found to promote the catalytic CO_2 cyclo-addition reaction. The MOF-892 exhibited high catalytic activity in the one-pot synthesis of styrene carbonate from styrene and CO_2, without requiring the preliminary synthesis and isolation of styrene oxide.

Mixed-linker metal-organic frameworks, consisting of various organic linkers functionalized by multiple functional groups, are anticipated to exhibit better carbon-capture, due to an increased surface heterogeneity. In order to explore this possibility, samples having 3 topologies (csq, ftw, scu) and with each one (figures 13 to 15) comprising one of 3 types of linker of differing length, were prepared[242] and functionalized with 3 types of functional group: –F, –NH_2 or –OCH_3. Monte Carlo simulations were used to estimate the CO_2 adsorption of all of the materials. Among the parent frameworks, which comprised identical building-blocks but in differing topologies, the ftw-version had the highest CO_2 working capacity and CO_2/N_2 selectivity. This was due to the strong affinity for CO_2. The CO_2 adsorption performance of the ftw-version exhibited an obvious dependence upon pore-size; consistent with the host-adsorbate interaction energy. The increased CO_2 adsorption which occurred upon functionalization also exhibited a pore-size dependence; especially in the case of –NH_2 functionalized frameworks in csq and scu topology. This was attributed mainly to an increased host-adsorbate Coulombic interaction.

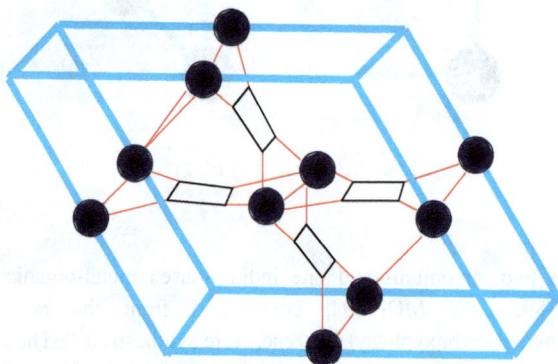

Figure 14. The csq topology

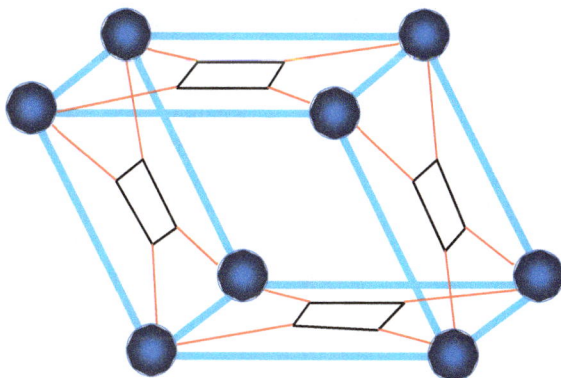

Figure 15. The scu topology

By using a conjugated amine-functionalized dicarboxylic ligand (H_2L = 2,2′-diamino-4,4′-stilbenedicarboxylic acid, $H_2SDCA-NH_2$), a porous and visible-light responsive zirconium-based metal-organic framework, $[Zr_6O_4(OH)_4(L)_6]\cdot 8DMF$, termed Zr-SDCA-$NH_2$, was prepared[243]. It exhibited good chemical stability, and broad visible-light absorption with an absorption edge at about 600nm. When used as a photocatalyst it exhibited visible-light activity for CO_2 reduction, with a formate-formation rate of 96.2μmol/h mmolMOF. The Zr_6 oxo cluster and the organic ligand both contributed to CO_2 photoreduction. A combination of amino groups and highly conjugated molecules appears to be a good route to designing a visible-light responsive photocatalyst.

A mesoporous zirconium-based metal-organic framework, JLU-MOF58, $[Zr_6O_4(OH)_8(H_2O)_4(TADIBA)_4]\cdot 24DMF\cdot 45H_2O$, where DMF is N,N-dimethylformamide and $H_2TADIBA$ is 4,4′-(2H-1,2,4-triazole-3,5-diyl) dibenzoic acid), was constructed[244]. It had a reo topology made up of bent ligands with Lewis-base sites and 8-connected Zr_6 clusters with Lewis and Brønsted acid sites. It comprised 2 types of mesoporous cage: octahedral and cuboctahedral (2.76 and 4.10nm), with a pore-volume of 1.76cm^3/g.

With the aim of capturing CO_2 at low pressures, due to the partial pressure (0.01 to 0.02MPa) of CO_2 in flue gases, a simple means was found[245] for preparing UiO-66 ($C_{34.94}H_{17.47}O_{61.74}Zr_6$, Zr^{IV} terephthalate) structures by mixing ligands of terephthalic acid and 1,2,4-benzenetricarboxylic acid. The free -COOH component of the 1,2,4-benzenetricarboxylic acid could be used to join deep eutectic solvents to the UiO-66, with reserved metal sites for CO_2 adsorption. As compared with 1,3,5-benzenetricarboxylic acid, the use of 1,2,4-benzenetricarboxylic acid maintained the stability of UiO-66 structure; this being more suitable for the capture of CO_2 from flue gases. Active -NH_2 and -OH groups, arising from additions, interacted with the intrinsic open metal sites so as to enhance CO_2 take-up at low pressures. This also improved the adsorption selectivity between CO_2 and N_2. The CO_2 take-up of the modified UiO-66 at 298K, was increased by 26% at 1bar and by 50% at 0.15bar; as compared with plain UiO-66. The CO_2/N_2 selectivity of the modified UiO-66 was 24.7 times higher than that of UiO-66 at low pressures. Strong interaction between the guest CO_2 and the modified UiO-66 was reflected by the existence of higher isosteric heats of adsorption in the modified UiO-66. Samples could be regenerated at 373K under vacuum and be recycled 6 times without a diminished CO_2 take-up.

Table 9. Catalytic tests of MIL-140A and MIL-140C
for the cyclo-addition of propylene oxide and CO_2

Catalyst	Conversion	Selectivity
ZrCl$_4$/1,4-benzenedicarboxylic acid/tetrabutylammoniumbromide	20%	>95%
ZrCl$_4$/biphenyl-4,4'-dicarboxylicacid/tetrabutylammoniumbromide	20%	>93%
tetrabutyl ammonium bromide	15%	>98%
tetrabutyl ammonium iodide	20%	>97%
MIL-140A/tetrabutyl ammonium bromide	82%	>99%
MIL-140C/tetrabutyl ammonium bromide	79%	>99%
MIL140A/tetrabutyl ammonium iodide	19%	>99%
MIL-140C/tetrabutyl ammonium iodide	10%	>99%

In other work[246], 2 very stable metal-organic frameworks which comprised chains of zirconium, coordinated with linkers of 1,4-H_2BDC (1,4-benzenedicarboxylic acid) [MIL-140A] and 4,4'-H_2BPDC (4,4'-biphenyldicarboxylic acid) [MIL-140C] were prepared and their catalytic activity was studied with regard to the coupling reaction of CO_2 with epoxides so as to produce cyclic carbonates under solvent-free conditions. Excellent activities were found in the case of both catalysts, and led to high epoxide conversion; with more than 99% selectivity in favour of cyclic carbonate. They were also entirely reusable after 4 cycles, with no appreciable loss in activity. The enhancement of catalytic activity was attributed to acidity/basicity effects. Propylene oxide was used as a substrate when determining the catalytic activity. At 80C, and a CO_2 pressure of 1.2MPa, there was no appreciable conversion of propylene oxide by either material alone within 6h. The presence of a co-catalyst such as tetrabutyl ammonium bromide led to a sharp increase in the propylene oxide conversion, to 79 to 82%, with a high propylene carbonate selectivity (table 9). The activity of tetrabutyl ammonium bromide was also tested. Under the experimental conditions of 80C and 1.2MPa of CO_2 for 6h, the conversion of propylene oxide using tetrabutyl ammonium bromide alone was much lower when compared with that observed using MIL-140A or MIL-140C systems. The conversion of propylene oxide was lower when compared with that produced by MIL-140/tetrabutyl ammonium bromide systems. The even distribution of active centers and the CO_2 adsorption ability of the metal-organic framework were suggested to be driving forces for improved catalysis. In order to check the versatility of MIL-140/tetrabutyl ammonium bromide catalysts, various epoxides were subjected to CO_2 cyclo-addition under identical reaction conditions (table 10). Terminal epoxides, such as allylglycidyl ether, propylene oxide and epichlorohydrin, were converted in appreciable quantities while cyclohexene oxide underwent very little conversion. This was attributed to steric hindrance.

Table 10. Performance of MIL-140A and MIL-140C
in the preparation of cyclic carbonates

Epoxide	Catalyst	Conversion (%)	Selectivity (%)
propylene oxide	MIL-140A	82	>99
epichlorohydrin	MIL-140[a]	80	>99
styrene oxide	MIL-140[a]	73	>98
cyclohexene oxide	MIL-140[a]	9	>93
allylglycidyl ether	MIL-140[a]	85	>97
propylene oxide	MIL-140C	79	>99
epichlorohydrin	MIL-140C	78	>99
styrene oxide	MIL-140C	69	>97
cyclohexene oxide	MIL-140C	7	>93
allylglycidyl ether	MIL-140C	84	>97

Another investigation[247] of the use of UiO-66 for CO_2/N_2 separation involved shaping the powder into pellets by using polyvinyl alcohol as a binder, and varying the ratio of metal-organic framework to the binder. An approximately 14% decrease in CO_2 adsorption capacity occurred during pelletization. A reduction in specific loading was observed, but the change in volumetric capacity was lower due to an increase in the bulk density during pelletization. The results of experiments which were performed on some 10g of UiO-66 in a column under 1.3bar at 300K revealed a preferential adsorption of CO_2 over N_2, and consequent CO_2 separation.

Table 11. CO_2 Adsorption/desorption properties of UiO-66-NH and polyethyleneimine-modified adsorbents

Absorbent	Condition	C_s (mmol$_{CO2}$/g)	α
UiO-66-NH$_2$	dry	2.67	25.5
UiO-66-NH$_2$	moist	2.13	25.5
polyethyleneimine^{c}48-UiO	dry	2.75	30.9
polyethyleneimine^{c}48-UiO	moist	2.69	30.9
polyethyleneimine^{c}72-UiO	dry	3.26	40.6
polyethyleneimine^{c}72-UiO	moist	3.43	40.6
polyethyleneimine72-UiO	dry	2.99	36.1
polyethyleneimine72-UiO	moist	3.02	36.1
polyethyleneimine^{c}96-UiO	dry	3.15	48.0
polyethyleneimine^{c}96-UiO	moist	3.33	48.0
polyethyleneimine96-UiO	dry	1.87	36.0
polyethyleneimine96-UiO	moist	2.69	36.0

α: adsorption selectivity of CO_2 over N_2, C_s: true CO_2 sorption capacity

In situ infra-red spectroscopy, combined with density functional theory, was used[248] to study the binding locations and energetics of CO_2 on UiO-66. Two unique CO_2 binding behaviors were deduced from the differing vibrational frequencies of the asymmetrical O-C-O stretching-mode. One configuration involved hydrogen bonding with μ_3-OH groups on the metal-organic framework nodes, while the other occurred when the CO_2 was stabilized by dispersive interactions. Adsorption enthalpies of -38.0 and -30.2kJ/mol were determined for the hydrogen-bonded and dispersion-stabilized complexes, respectively.

Samples of UiO-66-NH$_2$ were modified[249] by means of Schiff-base reactions between aldehyde groups in glutaraldehyde and amino groups in UiO-66-NH$_2$ and CO$_2$ pre-absorbed polyethyleneimine. As compared with plain UiO-66-NH$_2$, the resultant polyethyleneimine-modified metal-organic framework adsorbents, polyethyleneimine^C-UiO, exhibited reduced specific surface areas, of 7 to 150m^2/g, but retained the same crystal structure. The polyethyleneimine^{C}96-UiO adsorbent, in particular (table 11), had a markedly improved CO$_2$/N$_2$ adsorption selectivity, 48 as compared with 25, and a higher CO$_2$ adsorption capacity: 3.2 as compared with 2.7mmol/g. The adsorbent also had the moderate desorption energy of 68kJ/molCO$_2$.

Computational methods, involving density functional theory and force-field based molecular dynamics simulations, have been used[250] to predict the interfacial structures of metal-organic framework/polymer composites which consisted of UiO-66, combined with various polymers. The most compatible pairs were those which involved the more flexible polymers, such as polyvinylidene fluoride and polyethylene glycol. These exhibited closer contact at the surface, due to their greater ability to adapt their configurations to the surface of the metal-organic framework. Irregularities at the framework surface were filled by the polymer, and the terminal groups of the polymer could even penetrate into the pores of the metal-organic framework. In the case of polyethylene glycol, the pores of the framework could be sterically blocked due to strong framework/polymer interactions. Composites which involved polymers having a higher rigidity, such as intrinsic microporosity polymer-1 or polystyrene, thus had interfacial microvoids which decreased the surface contact between the two components. The most compatible composites involved polymers which had Young's moduli that were lower than 1GPa. Composites with polyvinylidene fluoride or polyethylene glycol were flexible continuous mixed-matrix membranes which possessed sufficient tensile strength and rigidity to yield useful films with a 70wt% metal-organic framework. Poor-compatibility systems, involving polystyrene or intrinsic microporosity polymer-1, exhibited interfacial microvoids which then increased the interaction distance between the framework and the polymer with this being reflected by brittle mixed-matrix membranes. Both the polyvinylidene fluoride and polyethylene glycol polymers penetrated the open pores of the metal-organic framework but, in the case of the latter, there was a stronger hydrogen-bonding between the framework and the polymer. This made the interface more compact and led to blocking of the framework pores. The polyvinylidene fluoride interacted strongly, but without pore-blocking. The polyvinylidene fluoride composites exhibited no change in rigidity as large amounts of metal-organic framework were incorporated, whereas the rigidity of the polyethylene glycol membranes increased by more than 100% upon incorporating the metal-organic framework. A comparison of the Young's modulus of pure polyethylene glycol film, with that of membranes involving 70wt% of UiO-66, showed that the latter were much more rigid; with the Young's modulus being increased from 133 to 284MPa. A similar comparison of pure polyvinylidene fluoride with the mixed-matrix membrane indicated a slight decrease, from 802 to 770MPa.

Ternary composites were prepared[251] by integrating CdS and molecular redox catalysts using the metal-organic framework, UiO-bpy. The CdS/UiO-bpy/Co composites were very effective in the photocatalytic conversion of CO_2 into CO using visible light, and the evolution-rate of CO could attain 235μmol/gh. This represented a more than 10-fold improvement over the performance of CdS, and the selectivity for CO was 85%. The outstanding performance of these composites was attributed to the separation and migration of photo-induced charge carriers, to an enhancement of the adsorption of CO_2 molecules and to the existence of abundant active sites for CO_2 reduction.

An aperture-opening process which results from dissociative linker exchange in zirconium-based, UiO-66, has been used[252] to encapsulate the ruthenium complex, (^{tBu}PNP)Ru(CO)HCl, within the framework, ^{tBu}PNP = 2,6-bis((di-tert-butyl-phosphino)methyl)pyridine. The resultant encapsulated complex was a very active catalyst for the hydrogenation of CO_2 to formate, could be recycled 5 times, showed no sign of bimolecular catalyst decomposition, and was not prone to poisoning.

A study of 8 functionalized zirconium-based materials showed[253] that non-functionalized UiO-66 resulted in the best (77%) conversion at low temperatures (50C), while the hydroxy-functionalized UiO-66-OH material gave the best (91%) conversion at high temperatures (140C). The material could be recycled up to 4 times without any appreciable decrease in reactivity.

On the basis of the compatibility principle of similar polymeric structures, and of steric interaction, polyether amine-modified UiO-66 was dispersed[254] in an ionic liquid having a polyether structure in order to form a new porous liquid at room temperature. This novel liquid exhibited a phenomenal CO_2 take-up capacity as compared with that of similar porous liquids.

Table 12. Properties of UiO metal-organic framework materials

MOF	Density (g/cm^3)	Pore Volume (cm^3/g)	Pore Surface Area (cm^2/g)
UiO-67	0.733	0.9435	2991
UiO-68	0.462	1.75	4189
UiO-69	0.348	2.4	4865
UiO-67-Li	0.778	0.8441	2689
UiO-67-Be	0.814	0.8263	2755
UiO-67-Ca	0.8785	0.7277	2440
UiO-67-Ti	0.8947	0.7416	2547
UiO-67-V	0.901	0.7390	2540
UiO-67-Mn	0.9093	0.7333	2542
UiO-67-Fe	0.911	0.7321	2515
UiO-68-Ti	0.546	1.46	3858
UiO-69-Ti	0.402	2.056	4492

Selective conversion of CO_2 to ethanol is of great interest but presents a significant challenge in forming a C–C bond while keeping a C–O bond intact throughout the process. Cooperative CuI sites on a Zr_{12} cluster of a metal–organic framework for selective hydrogenation of CO_2 to ethanol were reported[255]. With the assistance of an alkali cation, the spatially proximate Zr12-supported CuI centres activate hydrogen via bimetallic oxidative addition and promote C–C coupling to produce ethanol. The Cs^+-modified metal-organic framework catalyst, in 10h, produced ethanol with >99% selectivity and a turnover number (based upon all copper atoms) of 4080 in supercritical CO_2, with 30MPa of CO_2 and 5MPa of H_2 at 85C, or a turnover number of 490 at 2MPa of CO_2/H_2 (1/3) and 100C.

Defect-engineering has been used[256] to vary the band-gap of UiO-66 by adding various amino-functionalised benzoic acids at defective sites. This led to band-gaps in the range of 4.1 to 3.3eV, and first-principles calculations suggested that shrinkage of the band-gap was due to an upward shift in the valence-band energy, resulting from the presence of light-absorbing monocarboxylates. The photocatalytic properties of the defect-engineered materials with regard to the reduction of CO_2 to CO were improved.

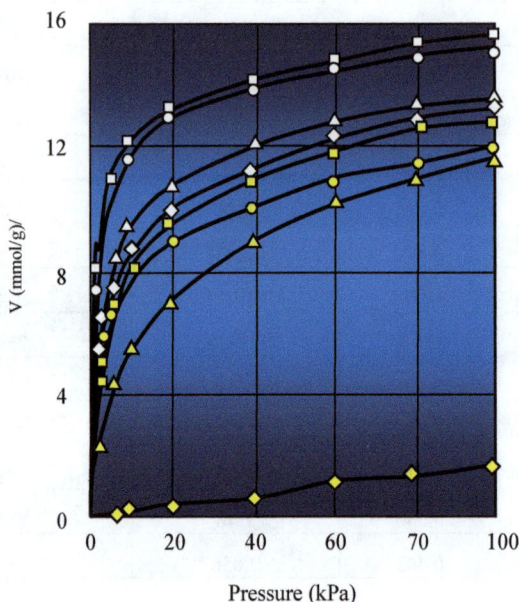

*Figure 16. Carbon dioxide take-up of metal alkoxide-functionalized UiO-67 at 298K.
Open squares: Ti, open circles: Ca, open triangles: Fe, open diamonds: Be, filled
squares: Mn, filled circles: V, filled triangles: Li, filled diamonds: non-functionalized*

Zirconium-based UiO-66 and UiO-66-NH$_2$ nanocrystals were developed[257] for the selective capture and removal of CO$_2$ from flue and natural gases. The UiO-66-NH$_2$ nanocrystals had a smaller grain size, contained a large number of defects and had -NH$_2$ groups within their pores. They exhibited selective CO$_2$ adsorptions, as compared with CH$_4$ and N$_2$, which amounted to theoretical separation-factors of 20 and 7, respectively. Monte Carlo simulations of the energies of CO$_2$, N$_2$ and CH$_4$ molecules in the gas mixture suggested that a much larger adsorption energy (0.32eV) of CO$_2$ over that of the other gases explained the selective adsorption.

The CO$_2$-adsorption ability of UiO-67 can be markedly changed[258] by incorporating alkali metals such as lithium, sodium and potassium, alkaline earths such as beryllium, magnesium and calcium and transition metals ranging from scandium to copper (table 12). The binding energy of CO$_2$ to beryllium, calcium, titanium, vanadium, manganese and iron alkoxide-functionalized ligands exceeds that of the lithium alkoxide-functionalized ligand. Monte Carlo simulations indicate clear CO_2 adsorption enhancement at 298K, and at pressures of up to 5000kPa, for these functionalized metal-organic framework materials; especially at low pressures. The titanium alkoxide-functionalized material exhibits the highest take-up amount (figure 16) of CO_2 at low pressures. The extension of organic linkers leads to a lower CO_2 adsorption capacity at low pressures because of a lower adsorption heat, but leads to a higher CO_2 adsorption capacity at high pressures because of an increase in pore volume.

Various metal-organic frameworks containing copper and zirconium clusters, and composites with graphene oxide (GO) were tested with regard to CO_2 adsorption[259]. The adsorption capacity and selectivity of samples of UiO-66, UiO-66-NH$_2$, HKUST-1 and the corresponding graphene oxide composites were compared at ambient temperatures. The incorporation of graphene oxide led to uniformly-shaped well-dispersed crystals having an increased pore volume due to additional porosity which formed in the interstices between the framework crystallites and the graphene oxide flakes. The UiO-66-NH$_2$ exhibited a CO_2 capacity of 3.07mmol/g at 25C and 4bar, and the highest CO_2/N_2 selectivity: 167 at 1bar. This was attributed to the amine functional groups in the organic linker, which increased the affinity with CO_2. With increasing pressure, the UiO-66-NH$_2$/GO composite exhibited a higher CO_2 capacity, and this was attributed to activation of the additional porosity between the framework crystallites and the graphene oxide layers. The HKUST-1/GO selectivity was increased, compared with that of HKUST-1, at all pressures.

Computational crystal-construction algorithms have been used[260] to create 12 metal-organic frameworks containing the newly synthesized [2,2'-bithiazole]-5,5'-dicarboxylic acid (H$_2$TzTz) spacer. Of the 12 structures, the material having the general formula, [Zr$_6$O$_4$(OH)$_4$(TzTz)$_6$], was the best candidate for carbon dioxide take-up and was subsequently synthesized. It was isoreticular to its bithiophene and bis(benzene) (UiO-67) analogues and crystallized in the cubic Pn$\bar{3}$ space group, featuring octahedral [Zr$_6$] nodes connected by 12 carboxylate groups from 6 bridging TzTz^{2-} spacers. It was largely microporous, with a micropore volume equal to 84% of the total pore volume. The Brunauer–Emmett–Teller area was 840m^2/g and it exhibited a maximum CO_2 take-up, at ambient pressures, of 2.3 and 1.7mmol/g at 273 and 298K, respectively. The CO_2 isosteric heat of adsorption was 18.7kJ/mol, and the CO_2/N_2 Henry selectivity was 10. The material was also tested as an heterogeneous catalyst for the reaction of CO_2 with epoxides having a CH$_2$X pendant arm, where X could be epichlorohydrin or epibromohydrin. This yielded the corresponding cyclic carbonates at 393K and P$_{CO2}$ =

1bar under solvent- and co-catalyst-free conditions. A conversion level of 74% was found, with epibromohydrin as a substrate.

Six-connected frameworks were prepared by adjusting the $ZrOCl_2 \bullet 8H_2O$/BTC molar ratio and were used[261] for the direct preparation of dimethyl carbonate from CO_2 and CH_3OH, with 1,1,1-trimethoxymethane being employed as a dehydration agent. It was found that the use of a suitable $ZrOCl_2 \bullet 8H_2O$/BTC molar ratio during synthesis was quite important in reducing the numbers of redundant BTC or zirconium clusters which were trapped in the micropores. Samples which had essentially no such redundant BTC or zirconium clusters had the greatest surface area, micropore size and number of acidic-basic sites. They therefore exhibited the best activity and the highest dimethyl carbonate yield (21.5%) under optimum reaction conditions. The present composition was even superior, due to the larger micropore size, than a material which had more acidic-basic sites and a greater surface area. This was largely because the larger micropore size allowed better access by the reactant to active sites located within the micropores.

Thermally and hydrolytically stable porous solids with zirconium- or hafnium-based metal-organic frameworks were synthesized. These materials, termed DUT, were[262] effective catalysts for both epoxide-CO_2 cyclo-addition reactions and for the catalytic transfer hydrogenation of ethyl levulinate. In particular, the 12-connected DUT-52(Zr) had a higher catalytic activity than that of 8- and 6-connected catalysts for the synthesis of cyclic carbonates as well as for the production of γ-valerolactone. The secondary building unit connectivity, the coexistence of a moderate number of acidic and basic sites, the Brunauer-Emmett-Teller surface area and the combined effects of the pores in the framework material appeared to affect the catalytic activity. The reaction mechanism of the DUT-52(Zr)-mediated cyclo-addition reaction of CO_2 and the catalytic transfer hydrogenation reactions were investigated by using periodic density functional theory calculations. Monte Carlo simulations predicted a strong interaction of CO_2 molecules with the DUT-52(Zr) framework. The DUT-series catalysts exhibited a remarkable tolerance for water. The DUT-52(Zr) was recyclable and was an efficient catalyst for cyclo-addition and catalytic transfer hydrogenation reactions on at least 5 occasions, with no obvious reduction in activity or structural integrity.

A sequential post-synthesis ionization and metalation method has been used[263] to prepare zirconium-based metal-organic frameworks which incorporated zinc porphyrin and imidazolium additives. Tetratopic [5,10,15,20-tetrakis (4-carboxyphenyl) porphyrinato] zinc^{II} (ZnTCPP) ligands could thereby be incorporated into the cationic zirconium-based metal-organic framework. These materials could be used for CO_2 capture and its conversion into cyclic carbonate, from allyl glycidyl ether and CO_2, under cocatalyst-free and 1bar conditions. The structural features and CO_2 affinity of these materials could also be modified by introducing imidazolium groups or doping the zinc sites. Three zirconium-oxo-cluster node and porphyrin-linker based metal-organic frameworks, MOF-525, PCN-222 and PCN-224, which exhibited linker connectivities of 12, 8 and 6 and were tested[264], together with their porphyrin-linker metalated analogues, as catalysts for CO_2 fixation via the cyclo-addition of CO_2 and propylene oxide to give propylene carbonate. The catalytic activity was generally related to the connectivity of the Zr-oxo nodes. The lowest (6-fold) connected material, PCN-224, exhibited the best catalytic activity of the series, while the more highly (8-fold and 12-fold) connected PCN-222 and MOF-525 materials were less active. The catalytic activity of the latter materials depended markedly upon defects. The 12-connected MOF-525 material had 16% of missing linker defects but exhibited a greater catalytic activity, as compared with that of the 8-connected PCN-222, which had fewer defects. The overall catalytic activity increased in dual-site catalysts when the porphyrin linkers were metalated with Mn^{III} and Zn^{II} centers which acted as additional Lewis acid sites. Water-resistant samples of the PCN-222 material were prepared[265] for the selective separation of CO_2/CH_4 and CO_2/N_2. At 298K, the CO_2 take-ups at 100 and 3000kPa were 1.16 and 13.67mmol/g, respectively. At 298K and 100kPa, the values of CO_2/CH_4 (50:50, v/v) and CO_2/N_2 (50:50, v/v) adsorption selectivity separately reached 4.3 and 73.7. At 298K and 3000kPa, the CO_2/CH_4 (50:50, v/v) and CO_2/N_2 (50:50, v/v) adsorption selectivities were 4.7 and 32.8, respectively. Breakthrough tests confirmed that the PCN-222 could effectively separate CO_2/CH_4 (both 50:50 and 10:90) and CO_2/N_2 (both 50:50 and 15:85) gas mixtures. The PCN-222 retained its original CO_2 adsorption capacity after 5 cycles of CO_2 adsorption-desorption.

$Rhodium^{III}$-porphyrin zirconium metal-organic frameworks have been prepared[266] via the self-assembly of a rhodium-based metalloporphyrin tetracarboxylic ligand, Rh(TCPP)Cl, where TCPP is tetrakis(4-carboxyphenyl)porphyrin), and $ZrCl_4$. This framework was stable at up to 270C and exhibited good chemical stability in a wide range of solvents, including water. Single crystals contained 3-dimensional channels (1.9nm x 1.9nm) and Rh-porphyrin units were exposed to the cavities. Calculations which were based upon N_2 adsorption at 77K indicated a Brunauer–Emmett–Teller surface area of 3015m^2/g. At 1atm, the CO_2 take-up was up to 42, 53 and 98cm^3/g at 308, 298 and 273K, respectively. Under visible light ($\geq$400nm), it could catalyze CO_2 reduction to formate ions with up to 99% selectivity. It could also be recycled 3 times. The composite catalyst, PCN-222(Fe)/C, with a 1:2 mass ratio exhibited[267] a high catalytic capability for the electrochemical conversion of CO_2 into CO at a 494mV overpotential with a current density of 1.2mA/cm^2 and a maximum efficiency of 91% in CO_2-saturated 0.5M $KHCO_3$ aqueous solution. The catalyst retained its crystallinity and stability after 10h of electrolysis at $-0.60V_{RHE}$, with an efficiency of 80.4%. It generated 334μmol of CO. Metalloporphyrins doped with rhodium or iridium, Rh-PMOF-1 and Ir-PMOF-1, are heterogeneous catalysts for the chemical fixation of CO_2 into cyclic carbonates, giving yields[268] of up to 99%. Particularly notable was the fact that these catalytic reactions proceed in low CO_2 concentrations, and that yields of 83 and 73% could be obtained by using 5%CO_2 in the presence of Rh-PMOF-1 and Ir-PMOF-1, respectively. The former material could be recycled 10 times with a negligible loss of catalytic activity.

The mixing of rigid and soft linkers in a metal-organic framework structure to achieve tunable structural flexibility was exemplified[269] by a series of stable isostructural Zr-MOFs built with natural C4 linkers (fumaric acid, succinic acid, and malic acid). As shown by the differences in linker bond stretching and bending freedom, these metal-organic frameworks displayed distinct responsive dynamics to external stimuli, such as changes in temperature or guest-molecule type. Comprehensive *in situ* characterizations reveal a clear correlation, between linker-nature and metal-organic framework dynamic behavior, which leads to the discovery of a multivariate flexible framework. It exhibits an optimum combination of good working capacity and a significantly enhanced selectivity for CO_2/N_2 separation.

Miscellaneous

In order to enhance the adsorption surface area and interaction energy for CO_2 capture, new frameworks originating from metal–organic framework were proposed by using[270] heterofullerene (C48B12) as a linker with and without lithium doping (C48B12-MOF-Li, C48B12-MOF) based upon density functional theory calculations and first-principles molecular dynamics simulations. Using Monte Carlo simulations, the separation of CO_2/CH_4 and CO_2/H_2 with different mixing ratios, and the adsorption capacity of CO_2 at low pressure was explored. Because of the introduction of metal sites, the lithium-modified framework, C48B12-MOF-Li, exhibited a sharp improvement in CO_2 capture and in selectivity for CO_2/CH_4 and CO_2/H_2 separation.

Samples of MOF-199 were synthesized[271] by using the solvent-thermal method, focusing upon the significant parameters (solvent, synthesis time, temperature) affecting the structure of the MOF-199. High crystallinity and a high surface area were found at a 1:1 ratio of the $H_2O:C_2H_5OH$, for solvent ratios ranging from 1:1 to 1:2. Varying the synthesis duration from 18 to 48h, and using temperatures ranging from 40 to 140C, showed that the best conditions were 100C and 48h. The appropriate activation condition was 60C and 16h. An attempt was made to synthesize a highly porous carbon adsorbent by carbonizing a crystalline metal–organic framework without any carbon precursors and focusing upon the adsorption of CO_2 and CH_4 gases and CO_2/CH_4 selectivity at 298, 323 and 348K using a volumetric apparatus. The metal-organic framework-derived porous carbon was prepared[272] by the direct carbonization of MOF-199 as a template at 900C under a nitrogen atmosphere. Amino-impregnated porous carbon samples exhibited enhanced adsorption capacities involving a combination of physical and chemical adsorption. Polyethyleneimine was selected as the amine source, which was found to greatly enhance CO_2 capture when supported on the porous carbon.

Isotope exchange techniques and the Sieverts method, microcalorimetry, breakthrough experiments and multi-scale modelling were used to investigate the thermodynamics and kinetics of equimolar CH_4/CO_2 mixture separation in metal-organic frameworks. The prototypical MOF-5 was selected as it allowed benchmarking of the present binary mixture results using reported pure gas data. For the first time, an experimental binary gas adsorption isotherm of CH_4/CO_2 on MOF-5 was reported[273] and compared with the respective pure gas isotherms. The equilibrium thermodynamic selectivity from isotope-exchange experiments for equimolar CH_4/CO_2 separation was 8.3, while a much lower value of 2.83 was obtained using ideal adsorption solution theory. The large standard deviation of the model selectivities and the significant deviation of averaged model selectivity from the experimental value clearly emphasised the need to determine the selectivity reliably via experiment. The kinetic selectivity for the binary mixture separation determined by combining the results of isotope-exchange with the linear driving force model was 0.73. The co-adsorption heats and excess take-up of both gases in mixtures were lower than those of pure gases; and the intrinsically weaker sorption of CH_4 on MOF-5 was further weakened by the presence of strongly interacting CO_2. Thermodynamic and kinetic selectivities and the co-adsorption heats quantitatively suggested that CH_4/CO_2 separation was driven by the equilibrium thermodynamic factors, with no significant contribution arising from kinetic factors.

A pore space partition approach has been used to create porous metal-organic frameworks, FJU-90, for C_2H_2/CO_2 separation under ambient conditions[274]. Due to the triangular ligand, 2,4,6-tris(4-pyridyl)pyridine, cylindrical channels in the original FJU-88 material were partitioned into uniformly interconnected pore cavities, thus leading to markedly reduced pore sizes: from 12.0Å x 9.4Å to 5.4Å x 5.1Å. Due to narrowing of the pore size, the activated FJU-90a took up a large volume of C_2H_2 (180cm^3/g), but rather less CO_2 (103cm^3/g), at 298K and 1bar. This rendered it the best porous metal-organic framework material for C_2H_2/CO_2 (50%:50%) separation.

The development of porous materials having a size-discriminatory effect for CO_2 and other molecules is complicated by the difficulty of controlling the pore size to within 3 to 4Å. A novel anion-pillared material embedded with so-called molecular rotors, for CO_2/CH_4 and CO_2/N_2 separation, has been proposed[275]. The temporarily fixed molecular rotors under ambient conditions provided a size-selective channel for CO_2 from CH_4 and N_2. There are only a few other examples which permit the molecular sieving of CO_2 and CH_4. Such a behavior gives ZU-66 a high separation selectivity for both CO_2/CH_4 (136) and CO_2/N_2 (355) and a high CO_2 capacity (4.56mmol/g, 298K, 1bar). Breakthrough experiments further suggested its potential for CO_2/CH_4 and CO_2/N_2 separation.

Photoresponsive metal–organic frameworks can be tailored for CO_2 adsorption by exploiting steric hindrance and structural changes due to weak interactions between the CO_2 and active sites[276]. The construction of materials having target-specific active sites is done by introducing tetraethylenepentamine into azobenzene-functionalized metal-organic frameworks. Amines are specific active sites for CO_2, and capture it selectively. Cis/trans isomerization of azobenzene components, triggered by UV/Vis light, can adjust the electrostatic potential of the amines enough to expose or shelter them and modulate the CO_2 adsorption at strong active sites.

Hydroxy metal-organic framework/polyimide mixed-matrix membranes also lead to high separation-performances with regard to CO_2 capture[277]. Mixed-matrix membranes can exceed Robeson upper bounds, with H_2 and CO_2 permeabilities of 907 and 650 barrer, respectively, and a CO_2/CH_4 selectivity of 32. This excellent performance resulted from interactions at the boundary of the hydroxy metal-organic framework and the carboxylic polymers involving strong hydrogen bonds.

The n-PrOH adsorption properties of the 1-dimensional-channel-like metal-organic framework InOF-1 revealed[278] a high affinity for this host-guest system owing to the shape of the adsorption isotherms at a low relative pressure, the presence of a hysteresis loop during the desorption process and a relatively high isosteric heat of adsorption. Monte Carlo simulations revealed that this thermodynamic behavior was related to a preferential interaction between n-PrOH and the μ_2-OH groups of the InOF-1 surface at low loading, whereas n-PrOH self-aggregated at higher guest concentrations to form firstly dimers and then clusters. The kinetics of n-PrOH were further characterized and the mobility of the guests was shown to be slow; probably due to the formation of the guest clusters. Unlike other solvents, n-PrOH confinement did not enhance CO_2 capture in InOF-1. Filling of the micropores of InOF-1 with CO_2 inhibited the adsorption of n-PrOH, and demonstrated an absence of over-solubility of n-PrOH in the presence of CO_2.

The adsorption of carbon dioxide on elastic layered metal-organic frameworks was investigated[279] during and after exposure to water. Two ELM variants, ELM-11 and ELM-12, were contacted with water vapor and the impact of cyclical exposure on the CO_2 capacity of the adsorbents was observed. ELM-11 was found to lose CO_2 capacity with each successive exposure to water, whereas ELM-12 retained CO_2 capacity over

Materials Research Forum LLC
https://doi.org/10.21741/9781644900857

four exposure cycles. Density functional theory calculations were performed to interpret these observations. Changing the counter-ion from the simple tetrafluoroborate (BF_4^-) to the larger and more complex trifluoromethanesulfonate ($CF_3SO_3^-$) anion expands the number of potential binding sites for adsorbate molecules. While CO_2 competes directly with other adsorbates for binding sites in ELM-11, CO_2 does not directly compete with other adsorbates in ELM-12 due to its preference for direct interaction with both fluorine and oxygen atoms in $CF_3SO_3^-$.

An L-aspartic acid based microporous metal-organic framework (MIP-202) was studied[280] for its adsorptive separation performance with regard to CO_2/CH_4 and CO_2/N_2 mixtures. Results showed that MIP-202 had ultra-high CO_2/CH_4 (72.9 and 241.5 for CO_2/CH_4 = 50/50 and 10/90, respectively) and CO_2/N_2 (1,950,000 and 2129.1 for CO_2/N_2 = 50/50 and 15/85, respectively) IAST selectivities at 298K and 100kPa. Metropolis Monte Carlo simulation calculations show that CO_2 with greater polarizability and quadruple moment tends to occupy the pore walls of large cages with higher polarity, while the less polar CH_4 or N_2 is largely adsorbed in the pore walls of small cages with lower polarity, resulting in the ultra-high CO_2/CH_4 and CO_2/N_2 separation selectivities. MIP-202 exhibits a low CO_2 adsorption enthalpy (17.2 to 30.7kJ/mol), superior persistent re-usability after 5-cycle adsorption experiments, easy desorption performance at 298K, moderate water and moisture stability, good stability in dry and humid SO_2 atmospheres, and low ligand cost ($36/kg). MIP-202 extrudates were prepared via an extrusion moulding method using hydroxypropyl cellulose as the binder and retained a 97.35% CO_2 take-up of MIP-202 powder at room temperature and pressure. This shows that MIP-202 is an industrially promising material for the separation of CO_2/CH_4 and CO_2/N_2 mixtures.

Six new metal-organic frameworks of Cu^{II}, Cd^{II} and Zn^{II} have been synthesized[281] from three flexible dicarboxylates and two rigid dicarboxylates together with 2-methyl-1-(4-(2-methyl-1H-imidazole-1-yl)butyl)-1H-imidazole as a co-ligand. All of these synthesized complexes have been characterized by single-crystal and powder X-ray diffraction and were further characterized by elemental analysis, infra-red spectroscopy and thermogravimetric analysis. Some N_2, CO_2, H_2 and CH_4 sorption studies were also performed for all of the metal-organic frameworks, and characteristic surface adsorptions were found in all cases. From studies of all of these complexes, a common rule-of-thumb is found: flexible dicarboxylates give a non-interpenetrating framework whereas rigid linkers give an interpenetrated framework.

A metal-organic framework with a highly ordered hierarchical structure was used[282] as an adsorbent and catalyst for the chemical fixation of CO_2 at atmospheric pressure. The CO_2 can be converted to the formate with excellent yields. Highly ordered macroporous and mesoporous structures were integrated into metal-organic frameworks, and the macro-, meso- and microporous structures have all been presented in one framework. Based upon the unique hierarchical pores, high surface area ($592m^2/g$), and high CO_2 adsorption capacity ($49.51cm^3/g$), the ordered macroporous-mesoporous metal-organic frameworks possessed a high activity for the chemical fixation of CO_2 (yield of 77%). These results provide a promising route to chemical CO_2 fixation through metal-organic framework materials.

A mono-alkyl pyrocarbonate-bipyridinium salt intermediate was used to activate CO_2 for synthesizing ethyl formate as a value-added product via tandem esterification and hydrogenation[283]. The required basic bipyridine sites for CO_2 activation and metallic Pd nanoparticles for hydrogenation were assembled in the same nano-cavity of a designer metal–organic framework. This metal-organic framework hybrid exhibited a high catalytic ability to generate ethyl formate (HCO_2Et) (1333μmol/gcat/h) with 93.5% selectivity at 412K. The turnover frequency of HCO_2Et, based upon the number of exposed palladium atoms, is up to 22.2/h; higher than those of other catalysts previously reported for generating alkyl formates from CO_2 under similar conditions.

Intracrystalline defects in metal-organic frameworks are known to play crucial roles in dictating their material properties. Computational simulations were used to induce linker vacancy defects into metal-organic framework membranes and to investigate their influence upon H_2/CH_4 separation[284]. Linker defective structures were created for the 228 candidate MOFs, and their separation performances were compared for defective and non-defective structures. The results showed that the existence of linker vacancy defects can lead to significant performance changes in the metal-organic framework membranes and, more importantly, the ranking of the best materials can differ as a function of the defects present.

Porous coordination polymers were synthesized[285] by using CO_2 and metal borohydrides as precursors. Borohydrides converted CO_2 into bridging ligands such as formate (HCO_2-) or formylhydroborate ($[BH(OCHO)_3]$) which are available to construct porous architectures; one of them having a $380m^2/g$ surface area.

A new strategy was proposed[286] for the preparation of mixed-matrix membranes with improved inorganic loadings plus enhanced compatibility and permeability. An *in situ* bottom-up growth approach was developed in which metal-organic framework precursors were well-dispersed within the polymerization mixture used for the preparation of the polymer membrane. Metal-organic framework precursors were then rearranged and uniformly distributed within a cross-linked hydrophilic PEG-based polymer membrane using thermal treatment at the melting temperature of the polymer. The resultant hybrid membranes contained up to 67.7wt% of metal-organic framework nanocrystals in dense and continuously ordered distributions within the polymer support. Compared with all of the reported metal-organic framework-based membranes, this value was currently the highest recorded. An excellent CO_2/N_2 selectivity of 38.5 and a CO_2 permeability of 1083.7 barrer were achieved, exceeding the known upper-bound limits defined for conventional polymer membranes.

Multifunctional nitrogen-doped nanoporous carbons prepared from metal-organic frameworks have been demonstrated. Metal-organic framework-derived nanoporous carbon possessed a porous structure having a surface area of 1244m^2/g and well-balanced pore sizes, as well as a 10% nitrogen content, which resulted[287] in good adsorption for CO_2 at a moderate storage pressure. Nitrogen-doped nanoporous carbon offered a CO_2 storage capacity of 10mmol/g at 45bar and 40C, and still maintained an adsorption of 7.2mmol/g at 100C. Due to the short pathway and enhancement of lithium-ion diffusion, it exhibited a high initial discharge capacity of 820mAh/g and a reversible charge capacity of 762mAh/g at a rate of 0.1A/g.

Hierarchical porosity and functionalization help to make full use of metal–organic frameworks, and a simple strategy has been reported for constructing hierarchically porous metal-organic frameworks through a competitive coordination method using tetrafluoroborate (M(BF4)$_x$, where M is a metal site) as both functional sites and etching agents[288]. The resultant metal-organic frameworks had *in situ* formed defect-mesopores and functional sites without sacrificing structural stability. The formation mechanism of the defect-mesopores was elucidated using a combination of experimental and first-principles calculations, indicating the general feasibility of the new approach. Compared with the original microporous counterparts, the new hierarchical metal-organic frameworks exhibited a superior adsorption of bulky dye molecules and their catalytic performance in CO_2 conversion was attributed to their specific hierarchical pore structures.

Ultra-thin metal-organic framework nanosheets show great potential in various separation applications. Metal-organic framework nanosheets are incorporated as a gutter layer in high-performance, flexible thin-film composite membranes for CO_2 separation. Ultra-thin (3 to 4nm) metal-organic framework nanosheets were prepared[289] using a surfactant-assisted method and subsequently coated onto a flexible porous support by vacuum-filtration. This produced an ultra-thin (~25nm), extremely flat metal-organic framework layer, which served as a highly permeable gutter with reduced gas resistance when compared with conventional polydimethylsiloxane gutter layers. Subsequent spin-coating of the ultra-thin metal-organic framework gutter layer with a polymeric selective layer (Polyactive) afforded a thin-film composite membrane exhibiting the best CO_2 separation performance then reported for a flexible composite membrane (CO_2 permeance of ~2100GPU with a CO_2/N_2 ideal selectivity of ~30). Several unique metal-organic framework nanosheets were examined as gutter layers, each differing with regard to structure and thickness (~10 and ~80nm); with results indicating that flexibility in the ultra-thin framework layer was critical for optimum membrane performance.

A class of metal-organic frameworks made from natural products has been grown[290] in an epitaxial fashion as films on the surfaces of glass substrates, which are modified with self-assembled monolayers of γ-cyclodextrin (γ-CD) molecules. The self-assembled monolayers are created by host-guest complexation of γ-CD molecules with surface-functionalized pyrene units. The CD-MOF films have a continuous polycrystalline morphology with a structurally out-of-plane (c-axial) orientation, covering an area of several square millimeters, with a thickness of about 2μm. Furthermore, this versatile host-guest strategy has been applied successfully in the growth of CD-MOFs as the shell on the curved surface of microparticles as well as in the integration of CD-MOF films into electrochemical devices for sensing carbon dioxide. In striking contrast to the control devices prepared from CD-MOF crystalline powders, these CD-MOF film-based devices display an enhancement in proton conductance of up to 300-fold. In addition, the CD-MOF film-based device exhibits more rapid and highly reversible CO_2-sensing cycles under ambient conditions, with a 50-fold decrease in conductivity upon exposure to CO_2 for 3s, which is recovered within 10s upon re-exposure to air.

The storage of CO_2 in the cavities of a porous coordination polymer was described using molecular rotor dynamics[291]. Due to the narrow pore windows of the porous coordination polymer, CO_2 was not adsorbed at 195K. As the temperature increased, the rotors exhibited rotational modes; such rotations dynamically expanded the size of the windows, leading to CO_2 adsorption. The rotational frequencies (10^{-6}s) of the rotors and the correlation times (10^{-8}s) of adsorbed CO_2 suggested that the slow rotation of the rotors sterically restricted CO_2 diffusion in the pores. This restriction results in an unusually slow CO_2 mobility close to the solid state. Once adsorbed at room temperature, the CO_2 was securely stored in the porous coordination polymer under vacuum at 195 to 233K because of steric hindrance of the rotors. This mechanism also applied to the storage of CH_4.

High-throughput computational screening has been used[292] to identify the top metal-organic framework membranes for flue gas separation. Grand canonical Monte Carlo and molecular dynamics simulations were used to assess the adsorption and diffusion properties of CO_2 and N_2 in 3806 different metal-organic frameworks. Using these data, selectivities and permeabilities of metal-organic framework membranes were predicted and compared with those of conventional membranes, polymers, and zeolites. The best-performing metal-organic framework membranes offering a CO_2/N_2 selectivity > 350 and CO_2 permeability > 106 barrer were identified. Ternary $CO_2/N_2/H_2O$ mixture simulations were then performed for the top MOFs in order to unlock their potential under industrial operating conditions, and the results showed that the presence of water decreases the CO_2/N_2 selectivity and CO_2 permeability of some MOF membranes. As a result of this stepwise screening procedure, the number of promising MOF membranes to be investigated for flue-gas separation in future experimental studies was narrowed down from thousands to tens. It was concluded that lanthanide-based MOFs with narrow pore openings (<4.5Å), low porosities (<0.75) and low surface areas (<1000m^2/g) are the best materials for membrane-based CO_2/N_2 separation.

Quantitative structure-property relationship models, with machine-learning, have been used to predict the CO_2 working capacity and CO_2/H_2 selectivity for carbon-capture using a topologically diverse database of 358400 hypothetical metal-organic framework structures and 1166 network topologies[293]. A tree- regression method permitted the use of 80% of the database as a training set, while the remainder was used for validation. The models were first built using purely geometric descriptors such as gravimetric surface area and void fraction. Additional models which accounted for chemical features of the frameworks were constructed using atomic property weighted radial distribution functions. The best models for CO_2 working capacity and CO_2/H_2 selectivity were built from a combination of 6 geometrical descriptors and 3 radial distribution function descriptors. The model could identify the 1000 best-performing metal-organic framework materials.

Computational screening of 6013 metal-organic frameworks for the simultaneous separation of H_2S and CO_2 from natural gas under humid conditions[294] was performed (represented by a six-component $CH_4/C_2H_6/C_3H_8/H_2S/CO_2/H_2O$ gas mixture). To minimize the competitive adsorption of H_2O, 606 hydrophobic MOFs are first selected on the basis of the Henry constants of H_2O and subsequently assessed for the adsorption capacity of H_2S + CO_2 (NH_2S+ CO_2) and the selectivity of H_2S + CO_2 over C1-C3 (SH_2S+ $CO_2/C1$-C3). Structure-performance relationships are established between MOF descriptors (the largest cavity diameter, surface area, void fraction and isosteric heat) and performance metrics (NH_2S+ CO_2 and SH_2S+ $CO_2/C1$-C3).

Hybrids with a nanosized metal-organic framework confined within a mesoporous structure have attracted increasing attention owing to their enhanced mass transfer and novel applications. However, effective control of MOF crystal growth within pores and further understanding of the structure-property relationship are challenging. The confinement of a nanosized metal-organic framework CAU-1 into a functionalized mesoporous polymer via a combined impregnation and solvent vapor growth process has been reported[295]. Carbonyl and hydroxyl groups, over the wall of the mesoporous polymer well, boost the nucleation and growth of CAU-1, leading to the formation of a nanosized MOF inside the mesoporous polymer. In contrast to bulk CAU-1, the nanosized CAU-1 within the hybrid exhibits a significantly enhanced CO_2 adsorption capacity at low pressures; as confirmed using ^{27}Al

References

[1]   Mohamedali, M., Henni, A., Ibrahim, H., Adsorption, 25[4] 2019, 675-692. https://doi.org/10.1007/s10450-019-00073-x

[2]   Ullah, S., Bustam, M.A., Assiri, M.A., Al-Sehemi, A.G., Sagir, M., Abdul Kareem, F.A., Elkhalifah, A.E.I., Mukhtar, A., Gonfa, G., Microporous and Mesoporous Materials, 288, 2019, 109569. https://doi.org/10.1016/j.micromeso.2019.109569

[3]   Yoo, D.K., Yoon, T.U., Bae, Y.S., Jhung, S.H., Chemical Engineering Journal, 380, 2020, 122496. https://doi.org/10.1016/j.cej.2019.122496

[4]   Vicent-Luna, J.M., Gutiérrez-Sevillano, J.J., Hamad, S., Anta, J., Calero, S., ACS Applied Materials and Interfaces, 10[35] 2018, 29694-29704. https://doi.org/10.1021/acsami.8b11842

[5]   Zhang, H., Li, J., Tan, Q., Lu, L., Wang, Z., Wu, G., Chemistry - a European Journal, 24[69] 2018, 18137-18157. https://doi.org/10.1002/chem.201803083

[6]   Guntern, Y.T., Pankhurst, J.R., Vávra, J., Mensi, M., Mantella, V., Schouwink, P., Buonsanti, R., Angewandte Chemie, 58[36] 2019, 12632-12639. https://doi.org/10.1002/anie.201905172

[7]   Sadeghi, N., Sharifnia, S., Do, T.O., Journal of Materials Chemistry A, 6[37] 2018, 18031-18035. https://doi.org/10.1039/C8TA07158F

[8]   Wang, X., Wisser, F.M., Canivet, J., Fontecave, M., Mellot-Draznieks, C., ChemSusChem, 11[18] 2018, 3315-3322. https://doi.org/10.1002/cssc.201801066

[9]   Ashling, C.W., Johnstone, D.N., Widmer, R.N., Hou, J., Collins, S.M., Sapnik, A.F., Bumstead, A.M., Midgley, P.A., Chater, P.A., Keen, D.A., Bennett, T.D., Journal of the American Chemical Society, 141[39] 2019, 15641-15648. https://doi.org/10.1021/jacs.9b07557

[10]   Mubashir, M., Yeong, Y.F., Chew, T.L., Lau, K.K., Industrial and Engineering Chemistry Research, 58[17] 2019, 7120-7130. https://doi.org/10.1021/acs.iecr.8b05773

[11]   Ntep, T.J.M.M., Wu, W., Breitzke, H., Schlüsener, C., Moll, B., Schmolke, L.,

Buntkowsky, G., Janiak, C., Australian Journal of Chemistry, 72[10] 2019, 835-841. https://doi.org/10.1071/CH19221

[12] Chang, Y.W., Chang, B.K., Journal of the Taiwan Institute of Chemical Engineers, 89, 2018, 224-233. https://doi.org/10.1016/j.jtice.2018.05.006

[13] Tarasov, A.L., Isaeva, V.I., Tkachenko, O.P., Chernyshev, V.V., Kustov, L.M., Fuel Processing Technology, 176, 2018, 101-106. https://doi.org/10.1016/j.fuproc.2018.03.016

[14] Liu, Q., Ding, Y., Liao, Q., Zhu, X., Wang, H., Yang, J., Colloids and Surfaces A, 579, 2019, 123645. https://doi.org/10.1016/j.colsurfa.2019.123645

[15] Nuhnen, A., Dietrich, D., Millan, S., Janiak, C., ACS Applied Materials and Interfaces, 10[39] 2018, 33589-33600. https://doi.org/10.1021/acsami.8b12938

[16] Chen, S., Mukherjee, S., Lucier, B.E.G., Guo, Y., Wong, Y.T.A., Terskikh, V.V., Zaworotko, M.J., Huang, Y., Journal of the American Chemical Society, 141[36] 2019, 14257-14271. https://doi.org/10.1021/jacs.9b06194

[17] Chen, S., Lucier, B.E.G., Luo, W., Xie, X., Feng, K., Chan, H., Terskikh, V.V., Sun, X., Sham, T.K., Workentin, M.S., Huang, Y., ACS Applied Materials and Interfaces, 10[36] 2018, 30296-30305. https://doi.org/10.1021/acsami.8b08496

[18] Zhang, J.W., Ji, W.J., Hu, M.C., Li, S.N., Jiang, Y.C., Zhang, X.M., Qu, P., Zhai, Q.G., Inorganic Chemistry Frontiers, 6[3] 2019, 813-819. https://doi.org/10.1039/C8QI01396A

[19] Abid, H.R., Rada, Z.H., Li, Y., Mohammed, H.A., Wang, Y., Wang, S., Arandiyan, H., Tan, X., Liu, S., RSC Advances, 10[14] 2020, 8130-8139. https://doi.org/10.1039/D0RA00305K

[20] Guan, L., Luo, G.H., Wang, Y., Jiegou Huaxue, 37[11] 2018, 1795-1804.

[21] Li, X.Y., Ma, L.N., Liu, Y., Hou, L., Wang, Y.Y., Zhu, Z., ACS Applied Materials and Interfaces, 10[13] 2018, 10965-10973. https://doi.org/10.1021/acsami.8b01291

[22] Zhang, E., Wang, T., Yu, K., Liu, J., Chen, W., Li, A., Rong, H., Lin, R., Ji, S., Zheng, X., Wang, Y., Zheng, L., Chen, C., Wang, D., Zhang, J., Li, Y., Journal of the American Chemical Society, 141[42] 2019, 16569-16573. https://doi.org/10.1021/jacs.9b08259

[23] Hu, L., Hong, X.J., Lin, X.M., Lin, J., Cheng, Q.X., Lokesh, B., Cai, Y.P., Crystal Growth and Design, 18[11] 2018, 7088-7093. https://doi.org/10.1021/acs.cgd.8b01234

[24] Zhou, H.L., Zhang, J.P., Chen, X.M., Frontiers in Chemistry, 6, 2018, 306.

[25] Li, Y.L., Zheng, Z., Nie, H., Zhao, C.M., Wang, Y.F., Li, J.J., Zhou, X.L., Zhang, J.H., E3S Web of Conferences, 118, 2019, 01044. https://doi.org/10.1051/e3sconf/201911801044

[26] Halder, A., Bhattacharya, B., Haque, F., Dinda, S., Ghoshal, D., Chemistry - a European Journal, 25[52] 2019, 12196-12205. https://doi.org/10.1002/chem.201902673

[27] Sadeghi, M., Isfahani, A.P., Shamsabadi, A.A., Favakeh, S., Soroush, M., Journal of Applied Polymer Science, 2019, 48704. https://doi.org/10.1002/app.48704

[28] Li, B., Inorganic Chemistry Communications, 88, 2018, 56-59. https://doi.org/10.1016/j.inoche.2017.12.014

[29] Wang, J., Huang, X.L., Cai, S.L., Zheng, S.R., Fan, J., Zhang, W.G., Polyhedron, 152, 2018, 17-21. https://doi.org/10.1016/j.poly.2018.06.025

[30] Li, Q., Luo, Y., Ding, Y., Wang, Y., Wang, Y., Du, H., Yuan, R., Bao, J., Fang, M., Wu, Y., Dalton Transactions, 48[24] 2019, 8678-8692. https://doi.org/10.1039/C9DT00478E

[31] Chen, L., Wang, Y., Yu, F., Shen, X., Duan, C., Journal of Materials Chemistry A, 7[18] 2019, 11355-11361. https://doi.org/10.1039/C9TA01840A

[32] Gładysiak, A., Deeg, K.S., Dovgaliuk, I., Chidambaram, A., Ordiz, K., Boyd, P.G., Moosavi, S.M., Ongari, D., Navarro, J.A.R., Smit, B., Stylianou, K.C., ACS Applied Materials and Interfaces, 10[42] 2018, 36144-36156. https://doi.org/10.1021/acsami.8b13362

[33] Alavijeh, R.K., Akhbari, K., White, J., Crystal Growth and Design, 19[12] 2019, 7290-7297. https://doi.org/10.1021/acs.cgd.9b01174

[34] Maity, R., Chakraborty, D., Nandi, S., Rinku, K., Vaidhyanathan, R., CrystEngComm, 20[39] 2018, 6088-6093. https://doi.org/10.1039/C8CE00752G

[35] Ntep, T.J.M.M., Reinsch, H., Liang, J., Janiak, C., Dalton Transactions, 48[42] 2019, 15849-15855. https://doi.org/10.1039/C9DT03518D

[36] D'Amato, R., Donnadio, A., Carta, M., Sangregorio, C., Tiana, D., Vivani, R., Taddei, M., Costantino, F., ACS Sustainable Chemistry and Engineering, 7[1] 2019, 394-402. https://doi.org/10.1021/acssuschemeng.8b03765

[37] Yulia, F., Nasruddin, Zulys, A., Ruliandini, R., Journal of Advanced Research in Fluid Mechanics and Thermal Sciences, 57[2] 2019, 158-174.

[38] Li, H., Wang, K., Hu, Z., Chen, Y.P., Verdegaal, W., Zhao, D., Zhou, H.C., Journal of Materials Chemistry A, 7[13] 2019, 7867-7874. https://doi.org/10.1039/C8TA11300A

[39]   Chen, C., Feng, N., Guo, Q., Li, Z., Li, X., Ding, J., Wang, L., Wan, H., Guan, G.,
        Journal of Colloid and Interface Science, 521, 2018, 91-101.
        https://doi.org/10.1016/j.jcis.2018.03.029

[40]   Khdhayyer, M., Bushell, A.F., Budd, P.M., Attfield, M.P., Jiang, D., Burrows,
        A.D., Esposito, E., Bernardo, P., Monteleone, M., Fuoco, A., Clarizia, G.,
        Bazzarelli, F., Gordano, A., Jansen, J.C., Separation and Purification Technology,
        212, 2019, 545-554. https://doi.org/10.1016/j.seppur.2018.11.055

[41]   Alivand, M.S., Shafiei-Alavijeh, M., Tehrani, N.H.M.H., Ghasemy, E., Rashidi,
        A., Fakhraie, S., Microporous and Mesoporous Materials, 279, 2019, 153-164.
        https://doi.org/10.1016/j.micromeso.2018.12.033

[42]   Guo, F., Yang, S., Liu, Y., Wang, P., Huang, J., Sun, W.Y., ACS Catalysis, 9[9]
        2019, 8464-8470. https://doi.org/10.1021/acscatal.9b02126

[43]   Shin, S., Yoo, D.K., Bae, Y.S., Jhung, S.H., Chemical Engineering Journal, 2019,
        123429. https://doi.org/10.1016/j.cej.2019.123429

[44]   Yoo, D.K., Khan, N.A., Jhung, S.H., Journal of CO2 Utilization, 28, 2018, 319-
        325. https://doi.org/10.1016/j.jcou.2018.10.012

[45]   Hu, Y.L., Zhang, R.L., Fang, D., Environmental Chemistry Letters, 17[1] 2019,
        501-508. https://doi.org/10.1007/s10311-018-0793-9

[46]   Chong, K.C., Lau, W.J., Lai, S.O., Thiam, H.S., Ismail, A.F., IOP Conference
        Series - Earth and Environmental Science, 268[1] 2019, 012010.
        https://doi.org/10.1088/1755-1315/268/1/012010

[47]   Liu, S., Liu, L.T., Sun, L.X., Zhou, Y.L., Xu, F., Polyhedron, 156, 2018, 195-199.
        https://doi.org/10.1016/j.poly.2018.09.033

[48]   Wang, L., Zhang, F., Wang, C., Li, Y., Yang, J., Li, L., Li, J., Separation and
        Purification Technology, 235, 2020, 116219.
        https://doi.org/10.1016/j.seppur.2019.116219

[49]   Ethiraj, J., Palla, S., Reinsch, H., Microporous and Mesoporous Materials, 294,
        2020, 109867. https://doi.org/10.1016/j.micromeso.2019.109867

[50]   Dhankhar, S.S., Nagaraja, C.M., New Journal of Chemistry, 43[5] 2019, 2163-
        2170. https://doi.org/10.1039/C8NJ04947E

[51]   Nath, K., Karan, C.K., Biradha, K., Crystal Growth and Design, 19[11] 2019,
        6672-6681. https://doi.org/10.1021/acs.cgd.9b01046

[52]   Ding, T., Zhang, S., Zhang, W., Zhang, G., Gao, Z.W., Dalton Transactions,
        48[22] 2019, 7938-7945. https://doi.org/10.1039/C9DT00510B

[53] Gupta, V., Mandal, S.K., Dalton Transactions, 48[2] 2019, 415-425.
https://doi.org/10.1039/C8DT03844A

[54] Mosca, N., Vismara, R., Fernandes, J.A., Tuci, G., Di Nicola, C., Domasevitch, K.V., Giacobbe, C., Giambastiani, G., Pettinari, C., Aragones-Anglada, M., Moghadam, P.Z., Fairen-Jimenez, D., Rossin, A., Galli, S., Chemistry - a European Journal, 24[50] 2018, 13170-13180.
https://doi.org/10.1002/chem.201802240

[55] Chen, D.M., Zhang, X.J., Journal of Solid State Chemistry, 278, 2019, 120906.
https://doi.org/10.1016/j.jssc.2019.120906

[56] Sikiti, P., Bezuidenhout, C.X., Van Heerden, D.P., Barbour, L.J., Inorganic Chemistry, 58[13] 2019, 8257-8262.
https://doi.org/10.1021/acs.inorgchem.9b00761

[57] Pal, A., Chand, S., Boquera, J.C., Lloret, F., Lin, J.B., Pal, S.C., Das, M.C., Inorganic Chemistry, 58[9] 2019, 6246-6256.
https://doi.org/10.1021/acs.inorgchem.9b00471

[58] Qin, B.W., Zhou, B.L., Cui, Z., Zhou, L., Zhang, X.Y., Li, W.L., Zhang, J.P., CrystEngComm, 21[10] 2019, 1564-1569. https://doi.org/10.1039/C8CE01942H

[59] Taylor, M.K., Runčevski, T., Oktawiec, J., Bachman, J.E., Siegelman, R.L., Jiang, H., Mason, J.A., Tarver, J.D., Long, J.R., Journal of the American Chemical Society, 140[32] 2018, 10324-10331. https://doi.org/10.1021/jacs.8b06062

[60] Chand, S., Pal, S.C., Mondal, M., Hota, S., Pal, A., Sahoo, R., Das, M.C., Crystal Growth and Design, 19[9] 2019, 5343-5353.
https://doi.org/10.1021/acs.cgd.9b00823

[61] Ji, X.H., Zhu, N.N., Ma, J.G., Cheng, P., Dalton Transactions, 47[6] 2018, 1768-1771. https://doi.org/10.1039/C7DT04882C

[62] Chand, S., Pal, A., Das, M.C., Chemistry - a European Journal, 24[22] 2018, 5982-5986. https://doi.org/10.1002/chem.201800693

[63] Zhao, X., Xu, H., Wang, X., Zheng, Z., Xu, Z., Ge, J., ACS Applied Materials and Interfaces, 10[17] 2018, 15096-15103. https://doi.org/10.1021/acsami.8b03561

[64] Song, X., Wu, Y., Pan, D., Zhang, J., Xu, S., Gao, L., Wei, R., Zhang, J., Xiao, G., Applied Catalysis A, 566, 2018, 44-51.
https://doi.org/10.1016/j.apcata.2018.08.011

[65] Ugale, B., Kumar, S., Dhilip Kumar, T.J., Nagaraja, C.M., Inorganic Chemistry, 58[6] 2019, 3925-3936. https://doi.org/10.1021/acs.inorgchem.8b03612

[66] Li, Y.P., Wang, Y., Xue, Y.Y., Li, H.P., Zhai, Q.G., Li, S.N., Jiang, Y.C., Hu,

M.C., Bu, X., Angewandte Chemie, 58[38] 2019, 13590-13595.
https://doi.org/10.1002/anie.201908378

[67]   Aulakh, D., Islamoglu, T., Bagundes, V.F., Varghese, J.R., Duell, K., Joy, M.,
        Teat, S.J., Farha, O.K., Wriedt, M., Chemistry of Materials, 30[22] 2018, 8332-
        8342. https://doi.org/10.1021/acs.chemmater.8b03885

[68]   Duan, X., Zheng, W., Yu, B., Ji, Z., Chemical Papers, 73[9] 2019, 2371-2375.
        https://doi.org/10.1007/s11696-019-00794-x

[69]   Duan, X., Yu, B., Lv, R., Ji, Z., Li, B., Cui, Y., Yang, Y., Qian, G., Polyhedron,
        155, 2018, 332-336. https://doi.org/10.1016/j.poly.2018.08.056

[70]   Kirandeep, Husain, A., Kharwar, A.K., Kataria, R., Kumar, G., Crystal Growth
        and Design, 19[3] 2019, 1640-1648. https://doi.org/10.1021/acs.cgd.8b01564

[71]   Parmar, B., Patel, P., Pillai, R.S., Tak, R.K., Kureshy, R.I., Khan, N.U.H., Suresh,
        E., Inorganic Chemistry, 58[15] 2019, 10084-10096.
        https://doi.org/10.1021/acs.inorgchem.9b01234

[72]   Meng, W., Zeng, Y., Liang, Z., Guo, W., Zhi, C., Wu, Y., Zhong, R., Qu, C., Zou,
        R., ChemSusChem, 11[21] 2018, 3751-3757.
        https://doi.org/10.1002/cssc.201801585

[73]   Lawson, S., Rownaghi, A.A., Rezaei, F., Energy Technology, 6[4] 2018, 694-701.
        https://doi.org/10.1002/ente.201700657

[74]   Ehrling, S., Senkovska, I., Bon, V., Evans, J.D., Petkov, P., Krupskaya, Y.,
        Kataev, V., Wulf, T., Krylov, A., Vtyurin, A., Krylova, S., Adichtchev, S.,
        Slyusareva, E., Weiss, M.S., Büchner, B., Heine, T., Kaskel, S., Journal of
        Materials Chemistry A, 7[37] 2019, 21459-21475.
        https://doi.org/10.1039/C9TA06781G

[75]   Liu, Y., Liu, S., Gonçalves, A.A.S., Jaroniec, M., RSC Advances, 8[62] 2018,
        35551-35556. https://doi.org/10.1039/C8RA07774F

[76]   Ge, B., Xu, Y., Zhao, H., Sun, H., Guo, Y., Wang, W., Materials, 11[8] 2018,
        1421. https://doi.org/10.3390/ma11081421

[77]   Majchrzak-Kucęba, I., Ściubidło, A., Journal of Thermal Analysis and
        Calorimetry, 138[6] 2019, 4139-4144. https://doi.org/10.1007/s10973-019-08314-
        5

[78]   Rani, P., Srivastava, R., Inorganic Chemistry Frontiers, 5[11] 2018, 2856-2867.
        https://doi.org/10.1039/C8QI00782A

[79]   Chang, M., Meng, X.X., Wang, Y., Zhang, W., International Journal of Hydrogen
        Energy, 44[56] 2019, 29583-29589.

https://doi.org/10.1016/j.ijhydene.2019.05.012

[80] Kurisingal, J.F., Rachuri, Y., Gu, Y., Kim, G.H., Park, D.W., Applied Catalysis A, 571, 2019, 1-11. https://doi.org/10.1016/j.apcata.2018.11.035

[81] Gargiulo, V., Alfè, M., Raganati, F., Lisi, L., Chirone, R., Ammendola, P., Fuel, 222, 2018, 319-326. https://doi.org/10.1016/j.fuel.2018.02.093

[82] Liu, Y., Ghimire, P., Jaroniec, M., Journal of Colloid and Interface Science, 535, 2019, 122-132. https://doi.org/10.1016/j.jcis.2018.09.086

[83] Szczęśniak, B., Choma, J., Microporous and Mesoporous Materials, 292, 2020, 109761. https://doi.org/10.1016/j.micromeso.2019.109761

[84] Shang, S., Tao, Z., Yang, C., Hanif, A., Li, L., Tsang, D.C.W., Gu, Q., Shang, J., Chemical Engineering Journal, 393, 2020, 124666. https://doi.org/10.1016/j.cej.2020.124666

[85] Mangal, S., Priya, S.S., Lewis, N.L., Jonnalagadda, S., Materials Today - Proceedings, 5[8] 2018, 16378-16389. https://doi.org/10.1016/j.matpr.2018.05.134

[86] Lu, X.T., Pu, Y.F., Li, L., Zhao, N., Wang, F., Xiao, F.K., Journal of Fuel Chemistry and Technology, 47[3] 2019, 338-343. https://doi.org/10.1016/S1872-5813(19)30016-7

[87] Pirzadeh, K., Ghoreyshi, A.A., Rahimnejad, M., Mohammadi, M., Korean Journal of Chemical Engineering, 35[4] 2018, 974-983. https://doi.org/10.1007/s11814-017-0340-6

[88] Qiu, Y.L., Zhong, H.X., Zhang, T.T., Xu, W.B., Su, P.P., Li, X.F., Zhang, H.M., ACS Applied Materials and Interfaces, 10[3] 2018, 2480-2489. https://doi.org/10.1021/acsami.7b15255

[89] Nam, D.H., Bushuyev, O.S., Li, J., De Luna, P., Seifitokaldani, A., Dinh, C.T., García De Arquer, F.P., Wang, Y., Liang, Z., Proppe, A.H., Tan, C.S., Todorović, P., Shekhah, O., Gabardo, C.M., Jo, J.W., Choi, J., Choi, M.J., Baek, S.W., Kim, J., Sinton, D., Kelley, S.O., Eddaoudi, M., Sargent, E.H., Journal of the American Chemical Society, 140[36] 2018, 11378-11386. https://doi.org/10.1021/jacs.8b06407

[90] Perfecto-Irigaray, M., Albo, J., Beobide, G., Castillo, O., Irabien, A., Pérez-Yáñez, S., RSC Advances, 8[38] 2018, 21092-21099. https://doi.org/10.1039/C8RA02676A

[91] Albo, J., Perfecto-Irigaray, M., Beobide, G., Irabien, A., Journal of CO2 Utilization, 33, 2019, 157-165. https://doi.org/10.1016/j.jcou.2019.05.025

[92] Ahlenhoff, K., Preischl, C., Swiderek, P., Marbach, H., Journal of Physical

Chemistry C, 122[46] 2018, 26658-26670.
https://doi.org/10.1021/acs.jpcc.8b06226

[93] Zhou, L., Niu, Z., Jin, X., Tang, L., Zhu, L., ChemistrySelect, 3[45] 2018, 12865-12870. https://doi.org/10.1002/slct.201803164

[94] Liu, S.M., Zhang, Z., Li, X., Jia, H., Ren, M., Liu, S., Advanced Materials Interfaces, 5[21] 2018, 1801062. https://doi.org/10.1002/admi.201801062

[95] Avci, G., Velioglu, S., Keskin, S., Journal of Physical Chemistry C, 123[46] 2019, 28255-28265. https://doi.org/10.1021/acs.jpcc.9b08581

[96] Chi, W.S., Sundell, B.J., Zhang, K., Harrigan, D.J., Hayden, S.C., Smith, Z.P., ChemSusChem, 12[11] 2019, 2355-2360.

[97] Zhang, Y., Zhang, Y., Wang, X., Yu, J., Ding, B., ACS Applied Materials and Interfaces, 10[40] 2018, 34802-34810. https://doi.org/10.1021/acsami.8b14197

[98] Mallick, A., Mouchaham, G., Bhatt, P.M., Liang, W., Belmabkhout, Y., Adil, K., Jamal, A., Eddaoudi, M., Industrial and Engineering Chemistry Research, 57[49] 2018, 16897-16902. https://doi.org/10.1021/acs.iecr.8b03937

[99] Wang, L., Jin, P., Huang, J., She, H., Wang, Q., ACS Sustainable Chemistry and Engineering, 7[18] 2019, 15660-15670.
https://doi.org/10.1021/acssuschemeng.9b03773

[100] Elsabawy, K.M., Fallatah, A.M., Journal of Inorganic and Organometallic Polymers and Materials, 28[6] 2018, 2865-2870. https://doi.org/10.1007/s10904-018-0913-9

[101] Murayama, T., Asano, M., Ohmura, T., Usuki, A., Yasui, T., Yamamoto, Y., Bulletin of the Chemical Society of Japan, 91[3] 2018, 383-390.
https://doi.org/10.1246/bcsj.20170371

[102] Kim, A.R., Yoon, T.U., Kim, S.I., Cho, K., Han, S.S., Bae, Y.S., Chemical Engineering Journal, 348, 2018, 135-142.
https://doi.org/10.1016/j.cej.2018.04.177

[103] Kim, M.K., Kim, H.J., Lim, H., Kwon, Y., Jeong, H.M., Electrochimica Acta, 306, 2019, 28-34. https://doi.org/10.1016/j.electacta.2019.03.101

[104] Franz, D.M., Dyott, Z.E., Forrest, K.A., Hogan, A., Pham, T., Space, B., Physical Chemistry Chemical Physics, 20[3] 2018, 1761-1777.
https://doi.org/10.1039/C7CP06885A

[105] Liu, J., Peng, L., Zhou, Y., Lv, L., Fu, J., Lin, J., Guay, D., Qiao, J., ACS Sustainable Chemistry and Engineering, 7[18] 2019, 15739-15746.
https://doi.org/10.1021/acssuschemeng.9b03892

[106] Ye, Y., Zhang, H., Chen, L., Chen, S., Lin, Q., Wei, F., Zhang, Z., Xiang, S., Inorganic Chemistry, 58[12] 2019, 7754-7759. https://doi.org/10.1021/acs.inorgchem.9b00182

[107] Wang, Z., Luo, X., Zheng, B., Huang, L., Hang, C., Jiao, Y., Cao, X., Zeng, W., Yun, R., European Journal of Inorganic Chemistry, 11, 2018, 1309-1314. https://doi.org/10.1002/ejic.201701404

[108] Zheng, B., Luo, X., Wang, Z., Zhang, S., Yun, R., Huang, L., Zeng, W., Liu, W., Inorganic Chemistry Frontiers, 5[9] 2018, 2355-2363. https://doi.org/10.1039/C8QI00662H

[109] Guo, W., Sun, X., Chen, C., Yang, D., Lu, L., Yang, Y., Han, B., Green Chemistry, 21[3] 2019, 503-508. https://doi.org/10.1039/C8GC03261K

[110] Huang, C., Dong, J., Sun, W., Xue, Z., Ma, J., Zheng, L., Liu, C., Li, X., Zhou, K., Qiao, X., Song, Q., Ma, W., Zhang, L., Lin, Z., Wang, T., Nature Communications, 10[1] 2019, 2779. https://doi.org/10.1038/s41467-019-10547-9

[111] Ashworth, D.J., Roseveare, T.M., Schneemann, A., Flint, M., Bernáldes, I.D., Vervoorts, P., Fischer, R.A., Brammer, L., Foster, J.A., Inorganic Chemistry, 58[16] 2019, 10837-10845. https://doi.org/10.1021/acs.inorgchem.9b01128

[112] Sensharma, D., Zhu, N., Tandon, S., Vaesen, S., Watson, G.W., Schmitt, W., Inorganic Chemistry, 58[15] 2019, 9766-9772. https://doi.org/10.1021/acs.inorgchem.9b00768

[113] Wei, L.Q., Ye, B.H., Inorganic Chemistry, 58[7] 2019, 4385-4393. https://doi.org/10.1021/acs.inorgchem.8b03525

[114] Gao, C.L., Zeitschrift fur Anorganische und Allgemeine Chemie, 644[16] 2018, 883-887.

[115] Xue, C., Wang, E., Feng, L., Yuan, Q., Hao, X., Xiao, B., Microporous and Mesoporous Materials, 264, 2018, 190-197. https://doi.org/10.1016/j.micromeso.2018.01.031

[116] Gong, J., Li, W., Li, S., Chinese Journal of Chemical Physics, 31[1] 2018, 52-60. https://doi.org/10.1063/1674-0068/31/cjcp1705108

[117] Wang, Y., He, M., Gao, X., Li, S., He, Y., Dalton Transactions, 47[21] 2018, 7213-7221. https://doi.org/10.1039/C8DT00863A

[118] Wen, H.M., Liao, C., Li, L., Alsalme, A., Alothman, Z., Krishna, R., Wu, H., Zhou, W., Hu, J., Chen, B., Journal of Materials Chemistry A, 7[7] 2019, 3128-3134. https://doi.org/10.1039/C8TA11596F

[119] Alduhaish, O., Lin, R.B., Wang, H., Li, B., Arman, H.D., Hu, T.L., Chen, B.,

Crystal Growth and Design, 18[8] 2018, 4522-4527.
https://doi.org/10.1021/acs.cgd.8b00506

[120] Jiang, J., Lu, Z., Zhang, M., Duan, J., Zhang, W., Pan, Y., Bai, J., Journal of the American Chemical Society, 140[51] 2018, 17825-17829.
https://doi.org/10.1021/jacs.8b07589

[121] Samarasinghe, S.A.S.C., Chuah, C.Y., Yang, Y., Bae, T.H., Journal of Membrane Science, 557, 2018, 30-37. https://doi.org/10.1016/j.memsci.2018.04.025

[122] Ansari, S.N., Kumar, P., Gupta, A.K., Mathur, P., Mobin, S.M., Inorganic Chemistry, 58[15] 2019, 9723-9732.
https://doi.org/10.1021/acs.inorgchem.9b00684

[123] Khan, J., Iqbal, N., Asghar, A., Noor, T., Materials Research Express, 6[10] 2019, 105539. https://doi.org/10.1088/2053-1591/ab3ff8

[124] Sabetghadam, A., Liu, X., Gottmer, S., Chu, L., Gascon, J., Kapteijn, F., Journal of Membrane Science, 570-571, 2019, 226-235.
https://doi.org/10.1016/j.memsci.2018.10.047

[125] He, X., Chen, D.R., Wang, W.N., Chemical Engineering Journal, 382, 2020, 122825. https://doi.org/10.1016/j.cej.2019.122825

[126] Zhang, Y., Yang, L., Wang, L., Duttwyler, S., Xing, H., Angewandte Chemie, 58[24] 2019, 8145-8150. https://doi.org/10.1002/anie.201903600

[127] Emerson, A.J., Knowles, G.P., Chaffee, A.L., Batten, S.R., Turner, D.R., CrystEngComm, 21[19] 2019, 3074-3085. https://doi.org/10.1039/C9CE00371A

[128] Yang, C.T., Kshirsagar, A.R., Eddin, A.C., Lin, L.C., Poloni, R., Chemistry - a European Journal, 24[57] 2018, 15167-15172.
https://doi.org/10.1002/chem.201804014

[129] Zha, J., Zhang, X., Crystal Growth and Design, 18[5] 2018, 3209-3214.
https://doi.org/10.1021/acs.cgd.8b00349

[130] Duan, X., Zhou, Y., Lv, R., Yu, B., Chen, H., Ji, Z., Cui, Y., Yang, Y., Qian, G., Journal of Solid State Chemistry, 260, 2018, 31-33.
https://doi.org/10.1016/j.jssc.2018.01.002

[131] Kochetygov, I., Bulut, S., Asgari, M., Queen, W.L., Dalton Transactions, 47[31] 2018, 10527-10535. https://doi.org/10.1039/C8DT01247D

[132] Nguyen, H.T.D., Tran, Y.B.N., Nguyen, H.N., Nguyen, T.C., Gándara, F., Nguyen, P.T.K., Inorganic Chemistry, 57[21] 2018, 13772-13782.
https://doi.org/10.1021/acs.inorgchem.8b02293

[133] Yan, Z.H., Du, M.H., Liu, J., Jin, S., Wang, C., Zhuang, G.L., Kong, X.J., Long, L.S., Zheng, L.S., Nature Communications, 9[1] 2018, 3353. https://doi.org/10.1038/s41467-018-05659-7

[134] Ntep, T.J.M.M., Reinsch, H., Schlüsener, C., Goldman, A., Breitzke, H., Moll, B., Schmolke, L., Buntkowsky, G., Janiak, C., Inorganic Chemistry, 58[16] 2019, 10965-10973. https://doi.org/10.1021/acs.inorgchem.9b01408

[135] Kobayashi, H., Taylor, J.M., Mitsuka, Y., Ogiwara, N., Yamamoto, T., Toriyama, T., Matsumura, S., Kitagawa, H., Chemical Science, 10[11] 2019, 3289-3294. https://doi.org/10.1039/C8SC05441J

[136] Zhu, J., Liu, J., Machain, Y., Bonnett, B., Lin, S., Cai, M., Kessinger, M.C., Usov, P.M., Xu, W., Senanayake, S.D., Troya, D., Esker, A.R., Morris, A.J., Journal of Materials Chemistry A, 6[44] 2018, 22195-22203. https://doi.org/10.1039/C8TA06383D

[137] Zhu, J., Usov, P.M., Xu, W., Celis-Salazar, P.J., Lin, S., Kessinger, M.C., Landaverde-Alvarado, C., Cai, M., May, A.M., Slebodnick, C., Zhu, D., Senanayake, S.D., Morris, A.J., Journal of the American Chemical Society, 140[3] 2018, 993-1003. https://doi.org/10.1021/jacs.7b10643

[138] Li, Y.H., Wang, S.L., Su, Y.C., Ko, B.T., Tsai, C.Y., Lin, C.H., Dalton Transactions, 47[28] 2018, 9474-9481. https://doi.org/10.1039/C8DT01405A

[139] Bratsos, I., Tampaxis, C., Spanopoulos, I., Demitri, N., Charalambopoulou, G., Vourloumis, D., Steriotis, T.A., Trikalitis, P.N., Inorganic Chemistry, 57[12] 2018, 7244-7251. https://doi.org/10.1021/acs.inorgchem.8b00910

[140] Zhang, Z.H., Wang, Q., Xue, D.X., Bai, J., Chemistry - an Asian Journal, 14[20] 2019, 3603-3610. https://doi.org/10.1002/asia.201900536

[141] Zou, L., Sun, X., Yuan, J., Li, G., Liu, Y., Inorganic Chemistry, 57[17] 2018, 10679-10684. https://doi.org/10.1021/acs.inorgchem.8b01330

[142] Rather, S.U., Muhammad, A., Al-Zahrani, A.A., Youssef, T.E., Petrov, L.A., Bulgarian Chemical Communications, 50[4] 2018, 608-614.

[143] Godfrey, H.G.W., Briggs, L., Han, X., Trenholme, W.J.F., Morris, C.G., Savage, M., Kimberley, L., Magdysyuk, O.V., Drakopoulos, M., Murray, C.A., Tang, C.C., Frogley, M.D., Cinque, G., Yang, S., Schröder, M., APL Materials, 7[11] 2019, 111104. https://doi.org/10.1063/1.5121644

[144] Zheng, J.J., Kusaka, S., Matsuda, R., Kitagawa, S., Sakaki, S., Journal of the American Chemical Society, 140[42] 2018, 13958-13969. https://doi.org/10.1021/jacs.8b09358

[145] Tang, M., Shen, H., Sun, Q., Journal of Physical Chemistry C, 123[43] 2019, 26460-26466. https://doi.org/10.1021/acs.jpcc.9b08359

[146] Nabais, A.R., Ribeiro, R.P.P.L., Mota, J.P.B., Alves, V.D., Esteves, I.A.A.C., Neves, L.A., Separation and Purification Technology, 202, 2018, 174-184. https://doi.org/10.1016/j.seppur.2018.03.028

[147] Song, X., Zhang, M., Chen, C., Duan, J., Zhang, W., Pan, Y., Bai, J., Journal of the American Chemical Society, 141[37] 2019, 14539-14543. https://doi.org/10.1021/jacs.9b07422

[148] Mao, F., Jin, Y.H., Liu, P.F., Yang, P., Zhang, L., Chen, L., Cao, X.M., Gu, J., Yang, H.G., Journal of Materials Chemistry A, 7[40] 2019, 23055-23063. https://doi.org/10.1039/C9TA07967J

[149] Fisher, D., Lead Halide Perovskites for Solar Cells, Materials Research Foundations,

[150] Wu, L.Y., Mu, Y.F., Guo, X.X., Zhang, W., Zhang, Z.M., Zhang, M., Lu, T.B., Angewandte Chemie, 58[28] 2019, 9491-9495. https://doi.org/10.1002/anie.201904537

[151] Sun, X., Wang, R., Ould-Chikh, S., Osadchii, D., Li, G., Aguilar, A., Hazemann, J.L., Kapteijn, F., Gascon, J., Journal of Catalysis, 378, 2019, 320-330. https://doi.org/10.1016/j.jcat.2019.09.013

[152] Jiang, S., Hu, Y., Chen, S., Huang, Y., Song, Y., Chemistry - a European Journal, 24[72] 2018, 19280-19288. https://doi.org/10.1002/chem.201804069

[153] Zhang, X., Chen, Z., Yang, X., Li, M., Chen, C., Zhang, N., Microporous and Mesoporous Materials, 258, 2018, 55-61. https://doi.org/10.1016/j.micromeso.2017.08.013

[154] Lee, W.R., Kim, J.E., Lee, S.J., Kang, M., Kang, D.W., Lee, H.Y., Hiremath, V., Seo, J.G., Jin, H., Moon, D., Cho, M., Jung, Y., Hong, C.S., ChemSusChem, 11[10] 2018, 1694-1707. https://doi.org/10.1002/cssc.201800363

[155] Xu, J., Liu, Y.M., Lipton, A.S., Ye, J., Hoatson, G.L., Milner, P.J., McDonald, T.M., Siegelman, R.L., Forse, A.C., Smit, B., Long, J.R., Reimer, J.A., Journal of Physical Chemistry Letters, 2019, 7044-7049. https://doi.org/10.1021/acs.jpclett.9b02883

[156] Choe, J.H., Kang, D.W., Kang, M., Kim, H., Park, J.R., Kim, D.W., Hong, C.S., Materials Chemistry Frontiers, 3[12] 2019, 2759-2767. https://doi.org/10.1039/C9QM00581A

[157] Pai, K.N., Baboolal, J.D., Sharp, D.A., Rajendran, A., Separation and Purification

Technology, 211, 2019, 540-550. https://doi.org/10.1016/j.seppur.2018.10.015

[158] Siegelman, R.L., Milner, P.J., Forse, A.C., Lee, J.H., Colwell, K.A., Neaton, J.B., Reimer, J.A., Weston, S.C., Long, J.R., Journal of the American Chemical Society, 141[33] 2019, 13171-13186. https://doi.org/10.1021/jacs.9b05567

[159] Abdelsayed, V., Gardner, T.H., Kababji, A.H., Fan, Y., Applied Catalysis A, 586, 2019, 117225. https://doi.org/10.1016/j.apcata.2019.117225

[160] Gholidoust, A., Maina, J.W., Merenda, A., Schütz, J.A., Kong, L., Hashisho, Z., Dumée, L.F., Separation and Purification Technology, 209, 2019, 571-579. https://doi.org/10.1016/j.seppur.2018.07.085

[161] Cheng, S., Wu, Y., Jin, J., Liu, J., Wu, D., Yang, G., Wang, Y.Y., Dalton Transactions, 48[22] 2019, 7612-7618. https://doi.org/10.1039/C9DT01249D

[162] Sapchenko, S.A., Barsukova, M.O., Belosludov, R.V., Kovalenko, K.A., Samsonenko, D.G., Poryvaev, A.S., Sheveleva, A.M., Fedin, M.V., Bogomyakov, A.S., Dybtsev, D.N., Schröder, M., Fedin, V.P., Inorganic Chemistry, 58[10] 2019, 6811-6820. https://doi.org/10.1021/acs.inorgchem.9b00016

[163] Feng, D.D., Zhao, Y.D., Wang, X.Q., Fang, D.D., Tang, J., Fan, L.M., Yang, J., Dalton Transactions, 48[29] 2019, 10892-10900. https://doi.org/10.1039/C9DT01430F

[164] Sharma, N., Dhankhar, S.S., Nagaraja, C.M., Microporous and Mesoporous Materials, 280, 2019, 372-378. https://doi.org/10.1016/j.micromeso.2019.02.026

[165] Sharma, N., Dhankhar, S.S., Kumar, S., Kumar, T.J.D., Nagaraja, C.M., Chemistry - a European Journal, 24[62] 2018, 16662-16669. https://doi.org/10.1002/chem.201803842

[166] Li, X.X., Liu, J., Zhang, L., Dong, L.Z., Xin, Z.F., Li, S.L., Huang-Fu, X.Q., Huang, K., Lan, Y.Q., ACS Applied Materials and Interfaces, 11[29] 2019, 25790-25795. https://doi.org/10.1021/acsami.9b03861

[167] Liu, S., Yao, S., Liu, B., Sun, X., Yuan, Y., Li, G., Zhang, L., Liu, Y., Dalton Transactions, 48[5] 2019, 1680-1685. https://doi.org/10.1039/C8DT04424D

[168] Li, Y.Z., Wang, H.H., Yang, H.Y., Hou, L., Wang, Y.Y., Zhu, Z., Chemistry - a European Journal, 24[4] 2018, 865-871. https://doi.org/10.1002/chem.201704027

[169] Fan, L., Kang, Z., Shen, Y., Wang, S., Zhao, H., Sun, H., Hu, X., Sun, H., Wang, R., Sun, D., Crystal Growth and Design, 18[8] 2018, 4365-4371. https://doi.org/10.1021/acs.cgd.8b00307

[170] Asgharnejad, L., Abbasi, A., Shakeri, A., Microporous and Mesoporous Materials, 262, 2018, 227-234. https://doi.org/10.1016/j.micromeso.2017.11.038

[171] Zhang, L., Jiang, K., Zhang, J., Pei, J., Shao, K., Cui, Y., Yang, Y., Li, B., Chen, B., Qian, G., ACS Sustainable Chemistry and Engineering, 7[1] 2019, 1667-1672. https://doi.org/10.1021/acssuschemeng.8b05431

[172] Jiang, M., Cui, X., Yang, L., Yang, Q., Zhang, Z., Yang, Y., Xing, H., Chemical Engineering Journal, 352, 2018, 803-810. https://doi.org/10.1016/j.cej.2018.07.104

[173] Singh, M., Solanki, P., Patel, P., Mondal, A., Neogi, S., Inorganic Chemistry, 58[12] 2019, 8100-8110. https://doi.org/10.1021/acs.inorgchem.9b00833

[174] Konavarapu, S.K., Ghosh, D., Dey, A., Pradhan, D., Biradha, K., Chemistry - a European Journal, 25[47] 2019, 11141-11146. https://doi.org/10.1002/chem.201902274

[175] Kurisingal, J.F., Babu, R., Kim, S.H., Li, Y.X., Chang, J.S., Cho, S.J., Park, D.W., Catalysis Science and Technology, 8[2] 2018, 591-600. https://doi.org/10.1039/C7CY02063E

[176] Li, J., Li, W.J., Xu, S.C., Li, B., Tang, Y., Lin, Z.F., Inorganic Chemistry Communications, 106, 2019, 70-75. https://doi.org/10.1016/j.inoche.2019.05.031

[177] Wang, L., Zou, R., Guo, W., Gao, S., Meng, W., Yang, J., Chen, X., Zou, R., Inorganic Chemistry Communications, 104, 2019, 78-82. https://doi.org/10.1016/j.inoche.2019.03.029

[178] Abdoli, Y., Razavian, M., Fatemi, S., Applied Organometallic Chemistry, 33[8] 2019, e5004. https://doi.org/10.1002/aoc.5004

[179] Wu, C., Irshad, F., Luo, M., Zhao, Y., Ma, X., Wang, S., ChemCatChem, 11[4] 2019, 1256-1263. https://doi.org/10.1002/cctc.201801701

[180] Wang, M., Liu, J., Guo, C., Gao, X., Gong, C., Wang, Y., Liu, B., Li, X., Gurzadyan, G.G., Sun, L., Journal of Materials Chemistry A, 6[11] 2018, 4768-4775. https://doi.org/10.1039/C8TA00154E

[181] Ugale, B., Dhankhar, S.S., Nagaraja, C.M., Crystal Growth and Design, 18[4] 2018, 2432-2440. https://doi.org/10.1021/acs.cgd.8b00065

[182] Chen, C., Zhang, M., Zhang, W., Bai, J., Inorganic Chemistry, 58[4] 2019, 2729-2735. https://doi.org/10.1021/acs.inorgchem.8b03308

[183] Zhang, G., Yang, H., Fei, H., ACS Catalysis, 8[3] 2018, 2519-2525. https://doi.org/10.1021/acscatal.7b04189

[184] Sun, Y., Han, H., Journal of Molecular Structure, 1194, 2019, 73-77.

[185] Jing, T., Chen, L., Jiang, F., Yang, Y., Zhou, K., Yu, M., Cao, Z., Li, S., Hong, M.,

Crystal Growth and Design, 18[5] 2018, 2956-2963.
https://doi.org/10.1021/acs.cgd.8b00068

[186] Li, N., Cai, Y., Shen, Q., Zhou, J., Journal of Photonics for Energy, 10[2] 2020, 023502.

[187] Zou, L., Yuan, J., Yuan, Y., Gu, J., Li, G., Zhang, L., Liu, Y., CrystEngComm, 21[21] 2019, 3289-3294. https://doi.org/10.1039/C9CE00343F

[188] Forse, A.C., Gonzalez, M.I., Siegelman, R.L., Witherspoon, V.J., Jawahery, S., Mercado, R., Milner, P.J., Martell, J.D., Smit, B., Blümich, B., Long, J.R., Reimer, J.A., Journal of the American Chemical Society, 140[5] 2018, 1663-1673.
https://doi.org/10.1021/jacs.7b09453

[189] Jin, C., Zhang, S., Zhang, Z., Chen, Y., Inorganic Chemistry, 57[4] 2018, 2169-2174. https://doi.org/10.1021/acs.inorgchem.7b03021

[190] Bien, C.E., Chen, K.K., Chien, S.C., Reiner, B.R., Lin, L.C., Wade, C.R., Ho, W.S.W., Journal of the American Chemical Society, 140[40] 2018, 12662-12666.
https://doi.org/10.1021/jacs.8b06109

[191] Han, L., Pham, T., Zhuo, M., Forrest, K.A., Suepaul, S., Space, B., Zaworotko, M.J., Shi, W., Chen, Y., Cheng, P., Zhang, Z., ACS Applied Materials and Interfaces, 11[26] 2019, 23192-23197. https://doi.org/10.1021/acsami.9b04619

[192] Wright, A.M., Wu, Z., Zhang, G., Mancuso, J.L., Comito, R.J., Day, R.W., Hendon, C.H., Miller, J.T., Dincă, M., Chem, 4[12] 2018, 2894-2901.
https://doi.org/10.1016/j.chempr.2018.09.011

[193] Ren, S., Feng, Y., Wen, H., Li, C., Sun, B., Cui, J., Jia, S., International Journal of Biological Macromolecules, 117, 2018, 189-198.
https://doi.org/10.1016/j.ijbiomac.2018.05.173

[194] Asadi, V., Kardanpour, R., Tangestaninejad, S., Moghadam, M., Mirkhani, V., Mohammadpoor-Baltork, I., RSC Advances, 9[49] 2019, 28460-28469.
https://doi.org/10.1039/C9RA04603H

[195] Chen, Y., Li, P., Noh, H., Kung, C.W., Buru, C.T., Wang, X., Zhang, X., Farha, O.K., Angewandte Chemie, 58[23] 2019, 7682-7686.
https://doi.org/10.1002/anie.201901981

[196] De Luna, P., Liang, W., Mallick, A., Shekhah, O., García De Arquer, F.P., Proppe, A.H., Todorović, P., Kelley, S.O., Sargent, E.H., Eddaoudi, M., ACS Applied Materials and Interfaces, 10[37] 2018, 31225-31232.
https://doi.org/10.1021/acsami.8b04848

[197] Zhu, D., Ao, S., Deng, H., Wang, M., Qin, C., Zhang, J., Jia, Y., Ye, P., Ni, H.,

ACS Applied Materials and Interfaces, 11[37] 2019, 33581-33588.
https://doi.org/10.1021/acsami.9b09811

[198] Pramchu, S., Jaroenjittichai, A.P., Laosiritaworn, Y., Greenhouse Gases: Science and Technology, 8[3] 2018, 580-586. https://doi.org/10.1002/ghg.1768

[199] Wang, Y., Cao, H., Zheng, B., Zhou, R., Duan, J., Crystal Growth and Design, 18[12] 2018, 7674-7682. https://doi.org/10.1021/acs.cgd.8b01433

[200] Liu, S., Dong, Q., Wang, D., Wang, Y., Wang, H., Huang, Y., Wang, S., Liu, L., Duan, J., Inorganic Chemistry, 58[23] 2019, 16241-16249.
https://doi.org/10.1021/acs.inorgchem.9b02774

[201] Hou, J., Wei, Y., Zhou, S., Wang, Y., Wang, H., Chemical Engineering Science, 182, 2018, 180-188. https://doi.org/10.1016/j.ces.2018.02.046

[202] Babu, D.J., He, G., Hao, J., Vahdat, M.T., Schouwink, P.A., Mensi, M., Agrawal, K.V., Advanced Materials, 31[28] 2019, 1900855.
https://doi.org/10.1002/adma.201900855

[203] Ren, Q., Yu, J.W., Luo, H.B., Zhang, J., Wang, L., Ren, X.M., Inorganic Chemistry, 58[21] 2019, 14693-14700.
https://doi.org/10.1021/acs.inorgchem.9b02358

[204] Zeeshan, M., Nozari, V., Yagci, M.B., Islk, T., Unal, U., Ortalan, V., Keskin, S., Uzun, A., Journal of the American Chemical Society, 140[32] 2018, 10113-10116.
https://doi.org/10.1021/jacs.8b05802

[205] Sun, L., Yun, Y., Sheng, H., Du, Y., Ding, Y., Wu, P., Li, P., Zhu, M., Journal of Materials Chemistry A, 6[31] 2018, 15371-15376.
https://doi.org/10.1039/C8TA04667K

[206] Lan, J., Liu, M., Lu, X., Zhang, X., Sun, J., ACS Sustainable Chemistry and Engineering, 6[7] 2018, 8727-8735.
https://doi.org/10.1021/acssuschemeng.8b01055

[207] Patel, P., Parmar, B., Kureshy, R.I., Khan, N.U., Suresh, E., ChemCatChem, 10[11] 2018, 2401-2408. https://doi.org/10.1002/cctc.201800137

[208] Abbasi, A.R., Moshtkob, A., Shahabadi, N., Masoomi, M.Y., Morsali, A., Ultrasonics Sonochemistry, 59, 2019, 104729.
https://doi.org/10.1016/j.ultsonch.2019.104729

[209] Wu, P., Li, Y., Zheng, J.J., Hosono, N., Otake, K.I., Wang, J., Liu, Y., Xia, L., Jiang, M., Sakaki, S., Kitagawa, S., Nature Communications, 10[1] 2019, 4362.
https://doi.org/10.1038/s41467-019-12414-z

[210] Gao, F.X., Ye, Y.J., Zhao, L.T., Liu, D.H., Li, Y., Inorganic Chemistry

Communications, 94, 2018, 39-42. https://doi.org/10.1016/j.inoche.2018.05.031

[211] Ullah, S., Bustam, M.A., Assiri, M.A., Al-Sehemi, A.G., Abdul Kareem, F.A., Mukhtar, A., Ayoub, M., Gonfa, G., Journal of Natural Gas Science and Engineering, 72, 2019, 103014. https://doi.org/10.1016/j.jngse.2019.103014

[212] Lysova, A.A., Samsonenko, D.G., Dorovatovskii, P.V., Lazarenko, V.A., Khrustalev, V.N., Kovalenko, K.A., Dybtsev, D.N., Fedin, V.P., Journal of the American Chemical Society, 141[43] 2019, 17260-17269. https://doi.org/10.1021/jacs.9b08322

[213] Kan, M.Y., Shin, J.H., Yang, C.T., Chang, C.K., Lee, L.W., Chen, B.H., Lu, K.L., Lee, J.S., Lin, L.C., Kang, D.Y., Chemistry of Materials, 31[18] 2019, 7666-7677. https://doi.org/10.1021/acs.chemmater.9b02539

[214] Röss-Ohlenroth, R., Bredenkötter, B., Volkmer, D., Organometallics, 38[18] 2019, 3444-3452. https://doi.org/10.1021/acs.organomet.9b00297

[215] He, H., Zhu, Q.Q., Guo, M.T., Zhou, Q.S., Chen, J., Li, C.P., Du, M., Crystal Growth and Design, 19[9] 2019, 5228-5236. https://doi.org/10.1021/acs.cgd.9b00621

[216] Rachuri, Y., Kurisingal, J.F., Chitumalla, R.K., Vuppala, S., Gu, Y., Jang, J., Choe, Y., Suresh, E., Park, D.W., Inorganic Chemistry, 58[17] 2019, 11389-11403. https://doi.org/10.1021/acs.inorgchem.9b00814

[217] Li, N., Liu, J., Liu, J.J., Dong, L.Z., Xin, Z.F., Teng, Y.L., Lan, Y.Q., Angewandte Chemie, 58[16] 2019, 5226-5231. https://doi.org/10.1002/anie.201814729

[218] Liu, J., Wei, Y., Bao, F., Li, G., Liu, H., Wang, H., Polyhedron, 169, 2019, 58-65. https://doi.org/10.1016/j.poly.2019.05.003

[219] Kim, S.H., Babu, R., Kim, D.W., Lee, W., Park, D.W., Chinese Journal of Catalysis, 39[8] 2018, 1311-1319. https://doi.org/10.1016/S1872-2067(17)63005-5

[220] Hungerford, J., Walton, K.S., Inorganic Chemistry, 58[12] 2019, 7690-7697. https://doi.org/10.1021/acs.inorgchem.8b03202

[221] Qiao, W., Song, T., Zhao, B., Chinese Journal of Chemistry, 37[5] 2019, 474-478. https://doi.org/10.1002/cjoc.201800587

[222] Uemura, K., Tomida, T., Yoshida, M., Journal of Solid State Chemistry, 270, 2019, 11-18. https://doi.org/10.1016/j.jssc.2018.10.039

[223] Zhang, Y.H., Bai, J., Chen, Y., Kong, X.J., He, T., Xie, L.H., Li, J.R., Polyhedron, 158, 2019, 283-289. https://doi.org/10.1016/j.poly.2018.10.067

[224] Sheng, D., Zhang, Y., Han, Y., Xu, G., Song, Q., Hu, Y., Liu, X., Shan, D., Cheng,

A., CrystEngComm, 21[24] 2019, 3679-3685.
https://doi.org/10.1039/C9CE00513G

[225] Zhai, Z.W., Yang, S.H., Lv, Y.R., Du, C.X., Li, L.K., Zang, S.Q., Dalton Transactions, 48[12] 2019, 4007-4014. https://doi.org/10.1039/C9DT00391F

[226] Zhu, Q.Q., Zhang, W.W., Zhang, H.W., Yuan, Y., Yuan, R., Sun, F., He, H., Inorganic Chemistry, 58[22] 2019, 15637-15643.
https://doi.org/10.1021/acs.inorgchem.9b02717

[227] Noh, J., Kim, D., Lee, J., Yoon, M., Park, M.H., Lee, K.M., Kim, Y., Kim, M., Catalysts, 8[11] 2018, 565. https://doi.org/10.3390/catal8110565

[228] Zhao, Y., Wang, L., Fan, N.N., Han, M.L., Yang, G.P., Ma, L.F., Crystal Growth and Design, 18[11] 2018, 7114-7121. https://doi.org/10.1021/acs.cgd.8b01290

[229] Cheng, A.L., Zhang, J., Ren, L.L., Gao, E.Q., Inorganica Chimica Acta, 482, 2018, 154-159. https://doi.org/10.1016/j.ica.2018.05.039

[230] Armaghan, M., Niu, R.J., Liu, Y., Zhang, W.H., Hor, T.S.A., Lang, J.P., Polyhedron, 153, 2018, 218-225. https://doi.org/10.1016/j.poly.2018.07.029

[231] Bolotov, V.A., Kovalenko, K.A., Samsonenko, D.G., Han, X., Zhang, X., Smith, G.L., McCormick, L.J., Teat, S.J., Yang, S., Lennox, M.J., Henley, A., Besley, E., Fedin, V.P., Dybtsev, D.N., Schröder, M., Inorganic Chemistry, 57[9] 2018, 5074-5082. https://doi.org/10.1021/acs.inorgchem.8b00138

[232] Sun, M.Y., Chen, D.M., Inorganic Chemistry Communications, 89, 2018, 18-21. https://doi.org/10.1016/j.inoche.2018.01.011

[233] Huang, N.Y., Mo, Z.W., Li, L.J., Xu, W.J., Zhou, H.L., Zhou, D.D., Liao, P.Q., Zhang, J.P., Chen, X.M., CrystEngComm, 20[39] 2018, 5969-5975.
https://doi.org/10.1039/C8CE00574E

[234] Khatua, S., Santra, A., Padmakumar, S., Tomar, K., Konar, S., ChemistrySelect, 3[2] 2018, 785-793. https://doi.org/10.1002/slct.201702975

[235] Li, Y., Liu, H., Wang, H., Qiu, J., Zhang, X., Chemical Science, 9[17] 2018, 4132-4141. https://doi.org/10.1039/C7SC04815G

[236] Vismara, R., Tuci, G., Mosca, N., Domasevitch, K.V., Di Nicola, C., Pettinari, C., Giambastiani, G., Galli, S., Rossin, A., Inorganic Chemistry Frontiers, 6[2] 2019, 533-545. https://doi.org/10.1039/C8QI00997J

[237] Sharma, N., Dhankhar, S.S., Nagaraja, C.M., Sustainable Energy and Fuels, 3[11] 2019, 2977-2982. https://doi.org/10.1039/C9SE00282K

[238] Dutta, G., Jana, A.K., Singh, D.K., Eswaramoorthy, M., Natarajan, S., Chemistry -

an Asian Journal, 13[18] 2018, 2677-2684. https://doi.org/10.1002/asia.201800815

[239] Sadeghi, N., Sharifnia, S., Do, T.O., Reaction Kinetics, Mechanisms and Catalysis, 125[1] 2018, 411-431. https://doi.org/10.1007/s11144-018-1407-z

[240] Ye, L., Gao, Y., Cao, S., Chen, H., Yao, Y., Hou, J., Sun, L., Applied Catalysis B, 227, 2018, 54-60. https://doi.org/10.1016/j.apcatb.2018.01.028

[241] Nguyen, P.T.K., Nguyen, H.T.D., Nguyen, H.N., Trickett, C.A., Ton, Q.T., Gutiérrez-Puebla, E., Monge, M.A., Cordova, K.E., Gándara, F., ACS Applied Materials and Interfaces, 10[1] 2018, 733-744. https://doi.org/10.1021/acsami.7b16163

[242] Li, W., Li, S., Chemical Engineering Science, 189, 2018, 65-74. https://doi.org/10.1016/j.ces.2018.05.042

[243] Sun, M., Yan, S., Sun, Y., Yang, X., Guo, Z., Du, J., Chen, D., Chen, P., Xing, H., Dalton Transactions, 47[3] 2018, 909-915. https://doi.org/10.1039/C7DT04062H

[244] Sun, X., Gu, J., Yuan, Y., Yu, C., Li, J., Shan, H., Li, G., Liu, Y., Inorganic Chemistry, 58[11] 2019, 7480-7487. https://doi.org/10.1021/acs.inorgchem.9b00701

[245] Li, Z., Sun, W., Chen, C., Guo, Q., Li, X., Gu, M., Feng, N., Ding, J., Wan, H., Guan, G., Applied Surface Science, 480, 2019, 770-778. https://doi.org/10.1016/j.apsusc.2019.03.030

[246] Jeong, H.M., Roshan, R., Babu, R., Kim, H.J., Park, D.W., Korean Journal of Chemical Engineering, 35[2] 2018, 438-444. https://doi.org/10.1007/s11814-017-0294-8

[247] Edubilli, S., Gumma, S., Separation and Purification Technology, 224, 2019, 85-94. https://doi.org/10.1016/j.seppur.2019.04.081

[248] Grissom, T.G., Driscoll, D.M., Troya, D., Sapienza, N.S., Usov, P.M., Morris, A.J., Morris, J.R., Journal of Physical Chemistry C, 123[22] 2019, 13731-13738. https://doi.org/10.1021/acs.jpcc.9b02513

[249] Zhu, J., Wu, L., Bu, Z., Jie, S., Li, B.G., ACS Omega, 4[2] 2019, 3188-3197. https://doi.org/10.1021/acsomega.8b02319

[250] Semino, R., Moreton, J.C., Ramsahye, N.A., Cohen, S.M., Maurin, G., Chemical Science, 9[2] 2018, 315-324. https://doi.org/10.1039/C7SC04152G

[251] Chen, C., Wu, T., Wu, H., Liu, H., Qian, Q., Liu, Z., Yang, G., Han, B., Chemical Science, 9[47] 2018, 8890-8894. https://doi.org/10.1039/C8SC02809E

[252] Li, Z., Rayder, T.M., Luo, L., Byers, J.A., Tsung, C.K., Journal of the American

Chemical Society, 140[26] 2018, 8082-8085. https://doi.org/10.1021/jacs.8b04047

[253] Noh, J., Kim, Y., Park, H., Lee, J., Yoon, M., Park, M.H., Kim, Y., Kim, M., Journal of Industrial and Engineering Chemistry, 64, 2018, 478-483. https://doi.org/10.1016/j.jiec.2018.04.010

[254] Zhao, X., Yuan, Y., Li, P., Song, Z., Ma, C., Pan, D., Wu, S., Ding, T., Guo, Z., Wang, N., Chemical Communications, 55[87] 2019, 13179-13182. https://doi.org/10.1039/C9CC07243H

[255] An, B., Li, Z., Song, Y., Zhang, J., Zeng, L., Wang, C., Lin, W., Nature Catalysis, 2[8] 2019, 709-717. https://doi.org/10.1038/s41929-019-0308-5

[256] Taddei, M., Schukraft, G.M., Warwick, M.E.A., Tiana, D., McPherson, M.J., Jones, D.R., Petit, C., Journal of Materials Chemistry A, 7[41] 2019, 23781-23786. https://doi.org/10.1039/C9TA05216J

[257] Li, P., Shen, Y., Wang, D., Chen, Y., Zhao, Y., Molecules, 24[9] 2019, 1822. https://doi.org/10.3390/molecules24091822

[258] Hu, J., Liu, Y., Liu, J., Gu, C., Journal of Physical Chemistry C, 122[33] 2018, 19015-19024. https://doi.org/10.1021/acs.jpcc.8b05334

[259] Pokhrel, J., Bhoria, N., Wu, C., Reddy, K.S.K., Margetis, H., Anastasiou, S., George, G., Mittal, V., Romanos, G., Karonis, D., Karanikolos, G.N., Journal of Solid State Chemistry, 266, 2018, 233-243. https://doi.org/10.1016/j.jssc.2018.07.022

[260] Müller, P., Bucior, B., Tuci, G., Luconi, L., Getzschmann, J., Kaskel, S., Snurr, R.Q., Giambastiani, G., Rossin, A., Molecular Systems Design and Engineering, 4[5] 2019, 1000-1013. https://doi.org/10.1039/C9ME00062C

[261] Xuan, K., Pu, Y., Li, F., Luo, J., Zhao, N., Xiao, F., Chinese Journal of Catalysis, 40[4] 2019, 553-566. https://doi.org/10.1016/S1872-2067(19)63291-2

[262] Kurisingal, J.F., Rachuri, Y., Palakkal, A.S., Pillai, R.S., Gu, Y., Choe, Y., Park, D.W., ACS Applied Materials and Interfaces, 11[44] 2019, 41458-41471. https://doi.org/10.1021/acsami.9b16834

[263] Liang, J., Xie, Y.Q., Wu, Q., Wang, X.Y., Liu, T.T., Li, H.F., Huang, Y.B., Cao, R., Inorganic Chemistry, 57[5] 2018, 2584-2593. https://doi.org/10.1021/acs.inorgchem.7b02983

[264] Epp, K., Semrau, A.L., Cokoja, M., Fischer, R.A., ChemCatChem, 10[16] 2018, 3506-3512. https://doi.org/10.1002/cctc.201800336

[265] Lv, D., Shi, R., Chen, Y., Chen, Y., Wu, H., Zhou, X., Xi, H., Li, Z., Xia, Q., Industrial and Engineering Chemistry Research, 57[36] 2018, 12215-12224.

https://doi.org/10.1021/acs.iecr.8b02596

[266] Liu, J., Fan, Y.Z., Li, X., Wei, Z., Xu, Y.W., Zhang, L., Su, C.Y., Applied Catalysis B, 231, 2018, 173-181. https://doi.org/10.1016/j.apcatb.2018.02.055

[267] Dong, B.X., Qian, S.L., Bu, F.Y., Wu, Y.C., Feng, L.G., Teng, Y.L., Liu, W.L., Li, Z.W., ACS Applied Energy Materials, 1[9] 2018, 4662-4669. https://doi.org/10.1021/acsaem.8b00797

[268] Liu, J., Fan, Y.Z., Li, X., Xu, Y.W., Zhang, L., Su, C.Y., ChemSusChem, 11[14] 2018, 2340-2347. https://doi.org/10.1002/cssc.201800896

[269] Wang, S., Xhaferaj, N., Wahiduzzaman, M., Oyekan, K., Li, X., Wei, K., Zheng, B., Tissot, A., Marrot, J., Shepard, W., Martineau-Corcos, C., Filinchuk, Y., Tan, K., Maurin, G., Serre, C., Journal of the American Chemical Society, 141[43] 2019, 17207-17216. https://doi.org/10.1021/jacs.9b07816

[270] Xu, G., Meng, Z., Liu, Y., Guo, X., Xiao, C., Deng, K., Microporous and Mesoporous Materials, 284, 2019, 385-392. https://doi.org/10.1016/j.micromeso.2019.04.046

[271] Alrubaye, R.T.A., Kareem, H.M., IOP Conference Series - Materials Science and Engineering, 557[1] 2019, 012060. https://doi.org/10.1088/1757-899X/557/1/012060

[272] Salehi, S., Anbia, M., Applied Organometallic Chemistry, 32[7] 2018, e4390. https://doi.org/10.1002/aoc.4390

[273] Zacharia, R., Gomez, L.F., Chahine, R., Cossement, D., Benard, P., Microporous and Mesoporous Materials, 263, 2018, 165-172. https://doi.org/10.1016/j.micromeso.2017.12.011

[274] Ye, Y., Ma, Z., Lin, R.B., Krishna, R., Zhou, W., Lin, Q., Zhang, Z., Xiang, S., Chen, B., Journal of the American Chemical Society, 141[9] 2019, 4130-4136. https://doi.org/10.1021/jacs.9b00232

[275] Yang, L., Cui, X., Zhang, Y., Wang, Q., Zhang, Z., Suo, X., Xing, H., ACS Sustainable Chemistry and Engineering, 7[3] 2019, 3138-3144. https://doi.org/10.1021/acssuschemeng.8b04916

[276] Jiang, Y., Tan, P., Qi, S.C., Liu, X.Q., Yan, J.H., Fan, F., Sun, L.B., Angewandte Chemie, 58[20] 2019, 6600-6604. https://doi.org/10.1002/anie.201900141

[277] Ma, C., Urban, J.J., ChemSusChem, 12[19] 2019, 4405-4411. https://doi.org/10.1002/cssc.201902248

[278] Álvarez, J.R., Mileo, P.G.M., Sánchez-González, E., Zárate, J.A., Rodríguez-Hernández, J., González-Zamora, E., Maurin, G., Ibarra, I.A., Journal of Physical

Chemistry C, 122[10] 2018, 5566-5577. https://doi.org/10.1021/acs.jpcc.8b00215

[279] Sotomayor, F.J., Lastoskie, C.M., Microporous and Mesoporous Materials, 292, 2020, 109371. https://doi.org/10.1016/j.micromeso.2019.03.019

[280] Lv, D., Chen, J., Yang, K., Wu, H., Chen, Y., Duan, C., Wu, Y., Xiao, J., Xi, H., Li, Z., Xia, Q., Chemical Engineering Journal, 375, 2019, 122074. https://doi.org/10.1016/j.cej.2019.122074

[281] Haque, F., Halder, A., Ghosh, S., Maiti, A., Ghoshal, D., Crystal Growth and Design, 19[9] 2019, 5152-5160. https://doi.org/10.1021/acs.cgd.9b00531

[282] Li, Z., Xing, X., Meng, D., Wang, Z., Xue, J., Wang, R., Chu, J., Li, M., Yang, Y., iScience, 15, 2019, 514-523. https://doi.org/10.1016/j.isci.2019.05.006

[283] An, B., Meng, Y., Li, Z., Hong, Y., Wang, T., Wang, S., Lin, J., Wang, C., Wan, S., Wang, Y., Lin, W., Journal of Catalysis, 373, 2019, 37-47. https://doi.org/10.1016/j.jcat.2019.03.008

[284] Kim, H., Lee, S., Kim, J., Langmuir, 35[11] 2019, 3917-3924. https://doi.org/10.1021/acs.langmuir.8b04175

[285] Kadota, K., Duong, N.T., Nishiyama, Y., Sivaniah, E., Horike, S., Chemical Communications, 55[63] 2019, 9283-9286. https://doi.org/10.1039/C9CC04771A

[286] Ma, L., Svec, F., Lv, Y., Tan, T., Journal of Materials Chemistry A, 7[35] 2019, 20293-20301. https://doi.org/10.1039/C9TA05401D

[287] Shi, X., Gong, J., Kierzek, K., Michalkiewicz, B., Zhang, S., Chu, P.K., Chen, X., Tang, T., Mijowska, E., New Journal of Chemistry, 43[26] 2019, 10405-10412. https://doi.org/10.1039/C9NJ01542F

[288] Chang, G.G., Ma, X.C., Zhang, Y.X., Wang, L.Y., Tian, G., Liu, J.W., Wu, J., Hu, Z.Y., Yang, X.Y., Chen, B., Advanced Materials, 2019, 1904969. https://doi.org/10.1002/adma.201904969

[289] Liu, M., Xie, K., Nothling, M.D., Gurr, P.A., Tan, S.S.L., Fu, Q., Webley, P.A., Qiao, G.G., ACS Nano, 12[11] 2018, 11591-11599. https://doi.org/10.1021/acsnano.8b06811

[290] Shen, D., Wang, G., Liu, Z., Li, P., Cai, K., Cheng, C., Shi, Y., Han, J.M., Kung, C.W., Gong, X., Guo, Q.H., Chen, H., Sue, A.C.H., Botros, Y.Y., Facchetti, A., Farha, O.K., Marks, T.J., Stoddart, J.F., Journal of the American Chemical Society, 140[36] 2018, 11402-11407. https://doi.org/10.1021/jacs.8b06609

[291] Inukai, M., Tamura, M., Horike, S., Higuchi, M., Kitagawa, S., Nakamura, K., Angewandte Chemie, 57[28] 2018, 8687-8690. https://doi.org/10.1002/anie.201805111

[292] Daglar, H., Keskin, S., Journal of Physical Chemistry C, 122[30] 2018, 17347-17357. https://doi.org/10.1021/acs.jpcc.8b05416

[293] Dureckova, H., Krykunov, M., Aghaji, M.Z., Woo, T.K., Journal of Physical Chemistry C, 123[7] 2019, 4133-4139. https://doi.org/10.1021/acs.jpcc.8b10644

[294] Qiao, Z., Xu, Q., Jiang, J., Journal of Materials Chemistry A, 6[39] 2018, 18898-18905. https://doi.org/10.1039/C8TA04939D

[295] Jiao, C., Majeed, Z., Wang, G.H., Jiang, H., Journal of Materials Chemistry A, 6[35] 2018, 17220-17226. https://doi.org/10.1039/C8TA05323E

Keyword Index

Pawley fitting, 40
pcu topology, 41
Pebax, 31, 64
pelletization, 94
permeability, 9, 15, 19, 31, 36, 43, 55, 64, 80, 106, 109
permeance, 44, 108
perovskite, 55
pervaporation, 75
photocatalysis, 2, 16
photoreduction, 7, 16, 37, 63, 70, 81, 89, 91
photosensitizer, 7, 68
physisorption, 29
plasmonic, 16
polar, 15, 22-23, 39, 41, 80, 105
Polyactive, 44, 107
polycatenated, 86
polymethacrylamide, 4
polymethyl methacrylate, 36
polysulfone, 8
porphyrin, 6, 16, 37, 55, 63, 88, 100-101
protonation, 36, 86
pseudotetrahedral, 51
pts topology, 51
pyrolysis, 29, 55

Qst, 17, 51, 88
quadrupole, 41

reo topology, 91
rhombic channel, 12, 88
rht topology, 38

Schläfli symbol, 15
scu topology, 90-91
seeded growth, 36, 80
selectivities, 7, 23-24, 28, 36-37, 45, 52, 63, 76, 84, 101, 103, 105, 108

Sieverts method, 103
sigmoidal, 54
solvothermal, 3, 5, 11-12, 15, 21-24, 27-28, 32-33, 47, 51, 53, 62, 64, 66-67, 71, 75, 79, 81-82, 85-86, 88
sorbates, 38
spirocyclic, 28
spiro-epoxy, 28
sql topology, 13, 15, 87
steric hindrance, 12, 28, 93, 104, 108
switchability, 29

templating, 24
tetracarboxylate, 11, 17, 51, 71
tetrahydrofuran, 8
tetrapodal, 22, 39
tetratopic, 21, 47
thermostable, 65
Torlon, 29
Toth, 79
trigonal-prismatic, 51
trinuclear, 51-52, 67
tritopic, 40, 53, 84
tunable, 38, 102

UiO, 2, 18, 32, 36, 48, 55, 91, 93-99
ultramicroporous, 27, 64, 73
unfunctionalized, 41
University of Oslo, 2

wettability, 62

zeolites, 1, 108
Zhejiang University, 2
zig-zag, 13
ZU, 2, 65, 104
zwitterionic, 27

About the Author

Dr. D.J. Fisher has wide knowledge and experience of the fields of engineering, metallurgy and solid-state physics, beginning with work at Rolls-Royce Aero Engines on turbine-blade research, related to the Concord supersonic passenger-aircraft project, which led to a BSc degree (1971) from the University of Wales. This was followed by theoretical and experimental work on the directional solidification of eutectic alloys having the ultimate aim of developing composite turbine blades. This work led to a doctoral degree (1978) from the Swiss Federal Institute of Technology (Lausanne). He then acted for many years as an editor of various academic journals, in particular *Defect and Diffusion Forum*. In recent years he has specialised in writing monographs which introduce readers to the most rapidly developing ideas in the fields of engineering, metallurgy and solid-state physics. His latest paper was published in *International Materials Reviews*, and he is co-author of the widely-cited student textbook, *Fundamentals of Solidification*.

www.ingramcontent.com/pod-product-compliance
Lightning Source LLC
Chambersburg PA
CBHW071703210326
41597CB00017B/2314